Venezuela

Venezuela
The Political Economy of Oil

Juan Carlos Boué

Published by the Oxford University Press
for the Oxford Institute for Energy Studies
1993

Oxford University Press, Walton Street, Oxford OX2 6DP
Oxford New York Toronto
Delhi Bombay Calcutta Madras Karachi
Kuala Lumpur Singapore Hong Kong Tokyo
Nairobi Dar es Salaam Cape Town
Melbourne Auckland Madrid

and associated companies in Berlin Ibadan

Oxford is a trade mark of Oxford University Press

British Library Cataloguing in Publication Data
available

ISBN 0-10-730012-X

Cover design by Moss, Davies, Dandy, Turner Ltd.
Typeset by Alden Multimedia
Printed in Great Britain on acid-free paper by
Bookcraft Ltd., Bath

The Political Economy of Oil-Exporting Countries

Venezuela: The Political Economy of Oil inaugurates a new series of books on the major petroleum exporting nations, most of them part of the developing world. These countries occupy a central position in the global economy given that oil is the energy source most used in the world and the most important primary commodity in international trade. At the same time they find themselves inescapably dependent on a single source of income. Their own economic prospects are closely bound to the future of their oil.

Books in this series incorporate research work done at the Oxford Institute for Energy Studies. Their aim is to provide a broad description of the oil and gas sectors of the country concerned, highlighting those features which give each country a physiognomy of its own. The analysis is set in the context of history, economic policy and international relations. It also seeks to identify the specific challenges that the exporting country studied will face in the future in developing its wealth to the best advantage of the economy.

CONTENTS

TABLES

FIGURES

ABBREVIATIONS

b/cd	barrels per calendar day
b/d	barrels per day
Btu	British Thermal Unit
cf	cubic feet
cf/d	cubic feet per day
cm^3	cubic centimetre
dwt	dead-weight tonnage
kW	kilowatt
m^3	cubic metre
m^3/d	cubic metres per day
mb	million barrels
mb/d	million barrels per day
mcf	million cubic feet
mcf/d	million cubic feet per day
mPa	millipascal second
mt	million tonnes
PdVSA	Petróleos de Venezuela S.A.
ppm	parts per million
tcf	trillion cubic feet

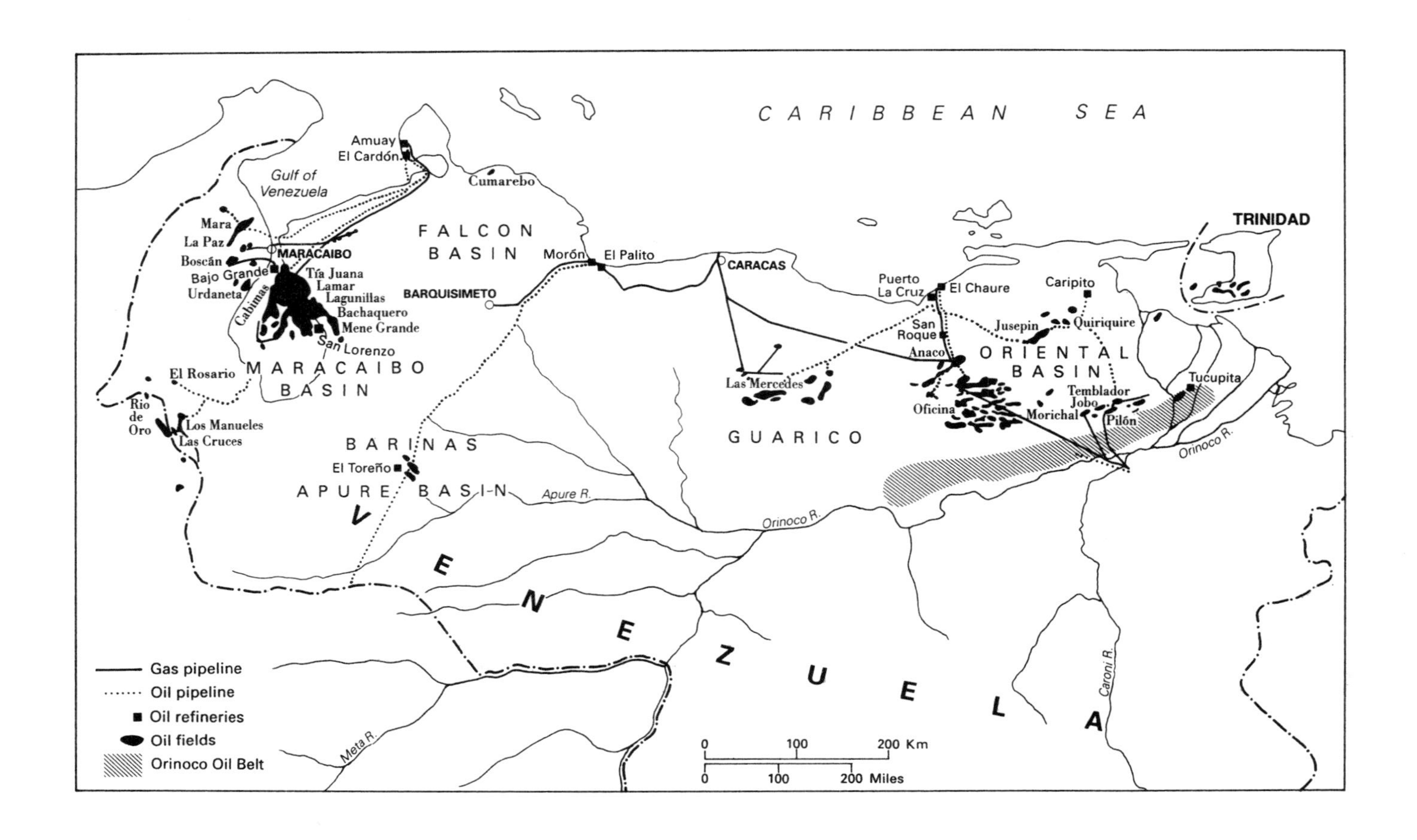

CARIBBEAN SEA
TRINIDAD
Amuay
El Cardón
Gulf of Venezuela
Cumarebo
FALCON BASIN
Mara
La Paz
MARACAIBO
Boscán
Bajo Grande
Urdaneta
Cabimas
Tía Juana
Lamar
Lagunillas
Bachaquero
Mene Grande
San Lorenzo
BARQUISIMETO
Morón
El Palito
CARACAS
Puerto La Cruz
El Chaure
Caripito
San Roque
Jusepin
Quiriquire
Anaco
ORIENTAL BASIN
Las Mercedes
Oficina
Temblador
Jobo
Morichal
Pilón
Tucupita
MARACAIBO BASIN
El Rosario
Rio de Oro
Los Manueles
Las Cruces
BARINAS
El Toreño
APURE BASIN
Apure R.
GUARICO
Orinoco R.
Caroni R.
Meta R.
VENEZUELA
Gas pipeline
Oil pipeline
Oil refineries
Oil fields
Orinoco Oil Belt
0
100
200 Km
0
100
200 Miles

1 INTRODUCTION

Venezuela has occupied a central place in the world petroleum market since the 1920s. Although the country is no longer producing more oil than the entire Middle East combined, as it did from 1926 to 1947, nevertheless, a serious student of the world oil market cannot afford to ignore Venezuela, because of three significant factors. First, despite the decline in its share of total world output, Venezuela is still a world class oil producer. Secondly, the national oil company, PdVSA, has grown to be the third largest oil company in the world,[1] and its forays into the downstream sector of the most important consumer nations have important implications for the security of oil supplies to developed nations as well as for the future of OPEC. Thirdly, Venezuela holds what could potentially be the biggest oil province in the world, the Orinoco Oil Belt. This area has not yet been extensively developed due to the extreme technological and economic challenges it poses but, clearly, a country that may have more oil in place than the whole of the Middle East combined has considerable potential importance.

This book attempts to describe, in as concise and analytical manner as possible, the most important geological, economic and political aspects of petroleum activities in Venezuela. This description is used as the touchstone to evaluate Caracas' future oil policies (as these result from economic, geological and political factors whose magnitude varies according to specific historical circumstances) and, in turn, to provide some insights into the prospects of the Venezuelan oil industry in the years to come. The study tries to dispel major uncertainties surrounding certain crucial aspects surrounding the Venezuelan upstream sector, such as the true nature and size of Venezuela's oil resource base, and the prospects of new oil finds in Venezuela's sedimentary basins, both on- and off-shore. It also addresses questions related to the development potential of the Orinoco Oil Belt as an oil province. Can Orinoco crude be sold in its virgin state as a refinery feedstock? If not, is the upgrading of this crude a feasible option under current and expected economic conditions? In this context, the interesting issue of the potentially revolutionary boiler fuel, Orimulsion, is addressed: does this fuel provide the only possible key to unlock the Orinoco's riches; or will its future be jeopardized by potentially tougher environmental restrictions in its target markets?

As regards development of the oil refining sector, the study focuses on the financial and environmental constraints that will define the decision arena for Venezuelan oil policy-makers during the foreseeable future. The Venezuelan refining system (domestic and overseas) will have to cope with the new environmental legislation in the advanced nations of the world and it will also have to adapt in order to process increasingly heavy feedstocks. This poses difficult problems which will be discussed. Finally, in the light of Venezuela's current economic situation, and the major changes in the international scene arising from the second Gulf war and the opening up of the oil sector of the former Soviet Union to the world economy, the study re-examines questions which have become topical. These relate to the role which the acquisitions of refining assets overseas may play in the expansion of the Venezuelan oil industry, the possible role and influence of foreign investment on the future of Venezuelan oil, and the country's membership of OPEC. These issues can generate policy responses which, by virtue of their potential 'demonstration effect' on other oil producers, could very well have significant implications for the structure of the world oil market in the future.

The book is divided into ten chapters. Chapter 1 is this introduction. Chapter 2 provides a brief historical and institutional overview of oil and gas activities in Venezuela, from the time of the first oil discoveries up to the present day. Chapters 3, 4, 5 and 6 deal with the main characteristics of the Venezuelan upstream and downstream sectors. The first focuses on the relevant features of the main hydrocarbons basins and oilfields in the country. The reserves position of the country is presented, as well as the main Venezuelan crude streams. Given the importance of the Orinoco Oil Belt, Chapter 4 is especially devoted to it, and to the revolutionary new fuel, Orimulsion, which PdVSA hopes will enable it to exploit the belt's vast resources. Chapter 5 analyses the development of natural gas in Venezuela. Finally, in Chapter 6, a detailed description of PdVSA's refining facilities is offered. In these chapters, the main problems of each sector of the oil industry are examined and discussed.

Chapters 7 and 8 are closely interrelated. The former deals with Venezuela's petroleum trade. The commercial structure (main markets, main type of exports) of the Venezuelan oil industry throughout history is discussed. The chapter also includes the main statistical traits of Venezuelan oil commerce: the quantity of crude oil and petroleum products exported, with a breakdown by type and market of destination; the statistics for the local demand of products, and so on. Chapter 8 examines the manner in which the country's dependence on one particular export market has led PdVSA to establish OPEC's biggest overseas refining empire, in a bid to minimize the volumetric uncertainty surrounding the country's oil exports. A full description of Venezuela's refining ventures abroad is provided, along with an explanation of the role which these ventures play, or are intended to play, in the wider scheme of the country's oil trading.

Chapter 9 studies the part which oil plays (and has played) in the Venezuelan economy. It relates how the country's fortunes have followed the vagaries of the oil market since the exploitation of the Lake Maracaibo field began, and ventures some suggestions as to where this relationship might be heading in the future. This chapter also presents the main facts about the Venezuelan fiscal regime concerning oil.

Finally, in Chapter 10 we draw together the strands presented in the various chapters, in order to offer an analysis of possible (and desirable) future developments for oil in Venezuela. In other words, having mentioned the main structural characteristics which define the problems, challenges and opportunities facing the Venezuelan oil industry, we will attempt – in the context of past experiences – to assess their significance and propose responses to them, because, as Leibnitz said, 'the present is loaded with the past, and pregnant with the future'.

Notes

1. Granted, according to criteria which give great weight to the size of a company's crude reserves. However, it is also the world's third largest refiner, after Exxon and Shell.

2 HISTORICAL BACKGROUND AND INSTITUTIONAL SET-UP

The purpose of this chapter is to provide the reader with some background information which will enable her or him to put in perspective our subsequent analysis of the Venezuelan oil industry. It focuses on six important elements in the history of oil in Venezuela. The first section describes the evolution of oil production in the country, which is described from the time of the first oil discoveries to the present day. The second section gives the legal framework which has regulated the operations of the oil industry in Venezuela. The third section concerns the oil companies holding concessions in the country prior to nationalization. The fourth section describes the institutional configuration of the national oil company, PdVSA, which started off as a disparate amalgamation of concessionary companies, but has turned into one of the most successful state oil companies. The fifth section characterizes the relationship existing between PdVSA and the ministry of energy and mines (which is what determines whether commercial or political considerations guide the country's oil policy). Finally, the sixth section concerns Venezuela's membership of the Organization of Petroleum Exporting Countries (OPEC).

2.1 Evolution of Venezuelan Oil Production

Petroleum was known to the aboriginal inhabitants of Venezuela long before the Spanish Conquest. The aborigines used asphalt for caulking boats and impregnating sails, pitch for lining handwoven reed baskets, and crude oil – collected from surface seepages (called *menes*) – for medicinal or illuminating purposes. When the Spaniards conquered Venezuela, they readily assimilated the Indian petroleum lore. For instance, the first mention of Venezuelan oil in a Spanish source - a reference to the oil seepages of Cubagua island, found in the *Historia General y Natural de las Indias, Islas y Tierra-Firme del Mar Océano*, written by Gonzalo Fernando de Oviedo y Valdés, and published in Seville in 1535 – credits oil as being a most excellent liquor and a very useful remedy for a plethora of sicknesses, including gout.[1] The Spanish royal family took the claims that oil was a miraculous elixir quite seriously, as is clear from a letter dated 1536 signed, quite aptly one might say, by queen Joanna the Mad, in which she ordered officials in Nueva Cádiz (Cubagua) to ship to her as much '*azeite petrolio*' as possible (Martínez, 1989, p. 4). A few years later, in 1539, the enterprising Don Francisco de Castellanos, treasurer of Nueva Cádiz, sent a barrel of Cubagua oil to King Charles I of Spain – Emperor Charles V of the Holy Roman Empire – in order to alleviate the terrible gout which beset the king (what effect the oil had on the royal personage's sickness is unknown). Cubagua, however, was not destined to become the first oil province of the world: on Christmas Day, 1541, an earthquake struck the island and razed Nueva Cádiz and all adjacent settlements to the ground. The Spaniards, however, found that other parts of Venezuela were also rich in oil. Unfortunately, the oil seepages in the vicinity of Lake Maracaibo proved to be especially useful not to them, but to seventeenth century French and English buccaneers, regular visitors of the Venezuelan shores, who took full advantage of the pitch found in these deposits to execute repairs on their weathered vessels in the lulls between plundering expeditions.

Commercial interest in Venezuelan oil began shortly after Edwin Drake discovered oil in Titusville. A concession to 'drill, produce and export petroleum or naphtha, or whatever the name is of the oil which exists in the earth [*sic*]' was granted in 1865, by General Jorge Sutherlan, president of Zulia state – where the Lake Maracaibo fields were later discovered – to Camillo Ferrand, a US citizen. Ferrand, however, neither drilled nor produced nor exported anything existing under the earth, probably because the Zulia legislature neglected to exempt him from paying duties on the imported machinery needed for the operations (Martínez, 1989, p. 15). A concession granted by the legislature of Nueva Andalucía state (since divided into Sucre and Monagas) to Manuel Olavarría did not fare much better. In 1883, however, seepage-oil production was established at La Alquitrana, near Cúcuta, Colombia. The roots of this operation could be traced to the earthquake which shook Colombia in 1875. The tremor was the strongest on record since 1610, and apart from property damage and loss of life, it caused enormous landslides on the slopes of the Andes. On the La Alquitrana plantation, after one of these landslides, oil began to seep from fissures, forming an iridescent film over the water in nearby streams and rivers. A group of six men decided to establish a company, the Compañía Petrolia del Táchira (see Martínez, 1986), to exploit these seepages. The company's operations were quite primitive, and on a very small scale: about 50 barrels per day (b/d) of crude from La Alquitrana (pumped from percussion-drilled holes, or scooped out of shallow pits) were carried to a primitive distillery located in the vicinity of the field, where it was refined in large vats heated by wood or charcoal fires. Kerosene the most valued product, was floated down river by barge to town markets in Colombia and Venezuela. Part of the residual oil was used as soap in sugar refining, as a herbicide, and as a wood preservative.

The event that led to the great Venezuelan oil boom, however, was not Petrolia's very modest commercial success, but the Mexican Revolution. For a number of years, after the discovery of the Golden Lane fields in Veracruz and Tamaulipas, Mexico had been the second most important oil exporter in the world. The Mexican Revolution, a decade-long civil war which began in 1910, reduced the country to rubble. Surprisingly, it left the oil industry in the country largely unscathed: Mexican production peaked at 500,000 b/d in 1921[2]. However, the instability associated with this event, as well as the rising nationalistic surge which was beginning to permeate Mexican politics, convinced the major oil companies that the time to look for greener pastures elsewhere had come. Thus, they moved to Venezuela, where the political scene under the dictatorship of Juan Vicente Gómez was perceived as more favourable to their interests. After an extremely difficult initial period, the search for oil in Venezuela began yielding fruits in 1914, when the Mene Grande field was discovered, to the east of Lake Maracaibo. Later, in 1917, Shell discovered the massive Bolívar Coastal Field. These finds, followed by the discovery of many other prolific fields in western and eastern Venezuela, vindicated those who had risked their reputations, their money, and even their lives, on the gamble of striking oil in this primitive and hostile country.[3] In 1929, Venezuela was producing 137 million barrels (mb) of crude, enough to make it the second most important country in the world, just after the USA, in terms of total oil output. It was also Royal Dutch/Shell's largest single source of production. In 1932, Venezuela assumed the enviable position of being the largest crude supplier for Great Britain – which in those days was easily the biggest oil importer in the world – ahead of the USA and Persia (Yergin, 1991, p. 236).

Venezuela's oil sector grew at a continuous pace from 1922 to 1935 (albeit interrupted by a brief episode of negative growth which lasted from 1931 to 1934, a consequence of the drop in oil demand caused by the great depression). After the worst of the depression was over, production expanded again until the year 1942. Then, in the midst of the second world war, when one would have expected a great surge in demand for Venezuelan oil, a transportation shortage caused by Shell's and Standard of New Jersey's decision to use their tankers to supply Britain from the USA, as well as by the effectiveness of Admiral Dönitz's U-Boat packs, forced Venezuelan oil production to its lowest levels since 1934. Once the war had finished, difficulties over energy supplies (caused by dislocations in important coal-producing areas of Europe and the USSR) fostered the growth of demand for energy from alternative supply points, such as Venezuela (Odell, 1986, p. 78). Furthermore, except for an ephemeral interlude of genuine democratic rule – the Acción Democrática party's first government, installed in February 1948 and overthrown in November of that same year – during which conditions in Venezuela ceased to be entirely to the companies' liking, the Venezuelan political environment of the 1950s proved highly auspicious for foreign investment. The dictatorship of Marcos Pérez Jiménez offered the major oil companies low taxes, a docile work force, non-existent government overseeing, a lifting of the ban on new concession rights, and high profits (Lieuwen, 1985, p. 203), and as a result, the majors ploughed their money prodigally into the country.[4] All of this 'ensured the continuation of the growth of the Venezuelan oil production and exports throughout the rest of the 1940s and up to 1957' (Odell, 1986, p. 78).

The main oil policy aim of the democratic governments which succeeded Pérez Jiménez in power after he was toppled by a pro-democracy military coup in 1958, was to increase the government's revenue coming from the industry. Steered by the clever and intensely nationalistic minister for mines and hydrocarbons, Juan Pablo Pérez Alfonzo, Venezuelan oil policy broke new ground in terms of the commercial relationships between the oil-producing countries and the major international oil companies. Especially important was the introduction in Venezuela of 'profit-sharing' arrangements. Profit sharing not only gave Venezuela the largest take per barrel produced of any of the major oil-exporting countries, but it also enabled the government 'to scrutinize information on both costs and prices, and to challenge the accounting procedures of the operating companies' (Danielsen, 1982, p. 141). Another important consequence of profit sharing was that it made it necessary for Venezuela to co-ordinate her oil policies with those of other oil producers, notably those of the Middle East. The reason for this was that Venezuela's tax regime put it at a disadvantage compared with these other producers, which had no profit-sharing arrangements. To negate this disadvantage, Venezuela established technical and political contacts with these countries, in an attempt to convince them of the wisdom of adopting taxing policies similar to its own. The Middle East producers followed Venezuela's lead, and the contacts at ministerial level which ensued as a result of this diplomatic offensive eventually led to the formation of OPEC.

The formation of OPEC was also a response to the unilateral reduction in posted prices enacted by the Seven Sisters in February 1959. This event prompted Juan Pablo Pérez Alfonzo to attend the First Arab Oil Congress – a brainchild of Saudi oil minister Abdullah Tariki – in order to convince the oil-exporting Arab nations that the best option to protect their interests as oil producers lay in placing less crude in the market.[5] In August 1960, when the oil companies dropped their posted prices once again, Venezuela, Saudi Arabia, Iran, Iraq and Kuwait responded by founding OPEC.

Resolution I-1 of the agreement establishing OPEC stated that 'members [of OPEC] could no longer remain indifferent to the attitude heretofore adopted by the oil companies in effecting price modifications.' Therefore, 'members shall demand that oil companies maintain their prices steady and free from all unnecessary modifications.' It warned that members would 'endeavour . . . to restore present prices to the level prevailing before the reductions', ensuring that, 'if any new circumstances arise which in the estimation of the oil companies necessitate price modifications, the . . . companies [would] enter into consultations with the . . . members affected

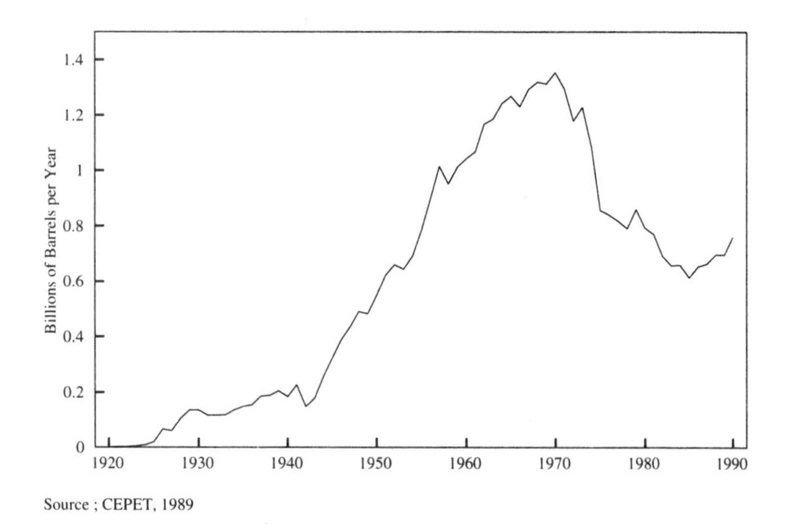

Figure 2.1: Historical Evolution of Venezuelan Oil Production. 1917–1990.

in order to explain the circumstances.' Most important of all, the resolution promised that OPEC members would 'study and formulate a system to ensure the stabilization of prices by . . . the regulation of production' (Evans, 1991, p. 395).

Throughout the 1960s, however, OPEC consistently failed to live up to Pérez Alfonzo's expectations. The desire of the Shah of Iran and the Arab rulers to achieve the highest level of production possible, as quickly as possible, killed off Pérez Alfonzo's attempts to introduce production discipline. Thus, the oil companies were able to maintain posted prices around $1.80 per barrel until the end of the decade. However, low prices were not the only troublesome prospect on the Venezuelan horizon: as a consequence of the country's higher production costs *vis-à-vis* the Middle East, the concessionaires dramatically reduced their investments in exploration, reserve expansion and equipment refurbishing from the early years in the decade onwards. Nevertheless, Venezuelan production continued to expand, reaching its maximum historical level in 1970 (3.7 mb/d). After this date, it began to decline steadily, and not even after the discovery of important new fields in both the eastern and western parts of the country was this trend reversed. It was only in 1990, when Iraq's invasion of Kuwait triggered a major crisis in the world oil market, that Venezuelan production once again enjoyed a positive growth rate. (Figure 2.1 shows the historical evolution of Venezuelan oil production since 1917.)

2.2 The Legal Framework of Oil in Venezuela

The first Venezuelan law dealing specifically with hydrocarbon exploration and exploitation was enacted in 1918.[6] It was the brainchild of Gumersindo Torres, Juan Vicente Gómez's minister for development. The law established a 15 per cent royalty on the notional export value of the crude produced (arrived at by applying a netback calculation based on the prevailing prices of refined products in the USA). Additionally, it set the maximum period for which territory could be held under a concession at 30 years, stipulated that drilling in a plot of land obtained under concession would have to begin no later than three years after exploration rights for the plot had been granted, and made provisions for setting aside half of each discovery as a national reserve. However, pressure from the oil companies persuaded Gómez to water down many of the provisions of Torres' law.[7] In this way, the Venezuelan upstream sector came to be ruled by the 'best petroleum law in the world – for the companies, that is' (Lieuwen, 1985, p. 197).[8]

In 1938, the government passed a law that would have given it higher royalties and taxes on concessions, and would also have altered the valuing system for Venezuelan crude and the relinquishment provisions in the concessions. The concessionaires objected strenuously to these revisions, and they threatened to take legal action against the Venezuelan government if the provisions were applied to their existing concessions.

The US government, however, with the Mexican nationalization still fresh in its memory, wanted to avoid a confrontation with the Venezuelans at all costs. Therefore, it assembled a team of experts to draft a new petroleum law acceptable to the Venezuelan government, and then proceeded to use strong-arm tactics with the companies, in order to persuade them to accept it. The resulting legislation was the 1943 Hydrocarbons Law. Under its provisions, royalties were raised to 16.66 per cent (as a standard rate for existing concessions and as a new rate for new concessions[9]) and the flat-rate taxes levied on each hectare of concession areas were increased. More importantly, the law established that at least 15 per cent of Venezuela's crude oil production had to be refined in the country.[10] Also, foreign oil companies were required for the first time to keep accounts of their local subsidiaries in Venezuela, and companies owning pipelines in the country were required to operate them on a common carrier basis (Evans, 1991, pp. 377–8).

The next major modifications to oil's legal status in Venezuela came after the consolidation of civilian rule in Venezuela (1958). The provisional military junta of 1958, which gave way to the government of Rómulo Betancourt, raised the maximum rate of progressive income tax to 45 per cent, from a level of 26 per cent. After this, oil minister Pérez Alfonzo boosted the government's take on company profits to about 67 per cent, a move that was greeted with dismay in oil company circles. Not content with this, the Acción Democrática government also stopped the granting of concessions, this time for good, and established a national oil company, CVP. Finally, from 1962 onwards, a government body called the Co-ordinating Commission for the Conservation and Marketing of Hydrocarbons was established to oversee the marketing aspect of the concessionaires' operations. The commission was given the power to demand compensation payments from the oil companies if they offered crude at a discount in excess of 10 per cent of the posted price for each grade.

Throughout the 1960s, the government gradually increased its take of oil company profits. At the same time, it introduced measures designed to strengthen its control over the pricing of crude oil for export purposes. Thus, in 1966, it began taxing companies with reference to 'artificial' export prices, which were invariably higher than the companies' realized prices.[11] Under the terms of the 1966 law, fixing the level of these reference prices was a matter involving both Venezuela and the oil companies, but as from March 1971, the government assumed exclusive control of this operation.

However, the government did not stop there. On the contrary, it sought to maximize its control over other aspects of the industry. In August 1971, President Rafael Caldera signed a law whereby the ownership of Venezuelan gas reverted back to the state. Caldera also signed the Hydrocarbons Reversion Law, which provided for the orderly return of all the concessions beginning in 1983. In June 1973, CVP was granted a monopoly of the domestic products market, and the Merchant Marine Law of the same year required that a specified proportion of Venezuelan production be carried in tankers flying the national flag from the moment the law became valid. In December 1972, Decree 832 established that the concessionaires operating in Venezuela had to submit, on an annual basis, their exploration, production, refining and export programmes to the government for approval. As a result of all these measures, says Coronel, 'for all practical purposes, the Venezuelan oil industry was in the hands of the state by 1972' (1983, p. 41). This might have been so, but the Venezuelan government, riding on the crest of the nationalist wave that swept the OPEC countries in the 1970s, sensed that the popular sentiment for early reversion had an irresistible momentum, and decided to proceed towards full nationalization.

As mentioned earlier, the nationalization law was promulgated in August 1975, and it became effective on 1 January, 1976.[12] Many Venezuelans then believed that their country, by taking into its hands the control of its most precious resource, had finally come of age. Sixteen years after the nationalization, however, the issue of equity participation by foreign oil companies in the Venezuelan oil industry is once again on the agenda. The wheel, it seems, has turned a full circle.

2.3 The Concessionaires

In 1913, Royal Dutch/Shell became the first major foreign oil company to have operations in Venezuela when it purchased a 51 per cent shareholding interest in the Caribbean Petroleum Company, which had exploration rights in 11 Venezuelan states. Caribbean (later brought under the control of a wholly-owned Shell subsidiary) was responsible for the Mene Grande and La Rosa (Cabimas) discoveries, which were instrumental in establishing Venezuela's oil sector. After the end of the first world war, American companies began following down the path that Shell had cleared, acquiring options and leases which covered large areas of Venezuela. Among these companies was the Creole syndicate (established in 1920), whose properties were producing 10 per cent of Venezuela's oil by 1927. The syndicate, however, had no refining outlets for its crude, or production expertise, for that matter. This made it dependent on Gulf Oil, which not only produced the oil in the concession areas on Creole's behalf, but also brought the crude to market after buying it from Creole under short-term buyback contracts. This state of affairs ended in 1928, when Standard Oil of New Jersey acquired a majority shareholding in the syndicate (which became the Creole Petroleum Corporation).

Another offshoot of the original Standard Oil trust which had extensive involvement in Venezuela during the early years of the oil industry there was Standard Oil of Indiana. For many years, its subsidiary company Lago was the most important producer of oil in the country and built an export refinery in the Dutch island of Aruba to process its Venezuelan output. In 1932, however, Creole purchased Lago's Venezuelan assets and concession rights, as well as the Aruba refinery. The deal was prompted by Indiana's fears that it would be unable to find outlets for its Venezuelan production in the USA, due to a new American tariff on imported oil. Jersey, thanks to its extensive overseas marketing empire, had no such worries. Eager to consolidate its position in Venezuela, Jersey gladly removed the burden from Indiana's shoulder.

Jersey's expansion did not end there, however. In 1937, through its subsidiary, the International Petroleum Company, it acquired half of the Mene Grande Oil Company's production operations. Mene Grande was owned by another of the Seven Sisters, Gulf Oil. In the early 1930s, Gulf – active in

Venezuela since the early 1920s, and producing 21 per cent of total output by 1929 – struck oil in Venezuela's eastern provinces. These finds were especially significant, because they consisted of reserves of light crude oil. Unfortunately for Gulf, it did not possess enough refining outlets to handle the increases in its equity production, and was therefore forced to divest part of Mene Grande's upstream operations. The deal was especially favourable for Jersey, since it gave International Petroleum the right to veto Mene Grande's exploration and development programme, and to establish a ceiling on the growth of its output relative to that of Creole. In 1938, Jersey sold half of its interest in Mene Grande to the Nederlandse Olie Maatschappij, a subsidiary of the Shell Group. Thereafter, Mene Grande's production expenditures were always divided between Gulf, Jersey and Shell (in the proportions of 50 per cent, 25 per cent and 25 per cent, respectively). With the Mene Grande deal, Jersey became the largest operator in Venezuela, accounting for 52 per cent of the total production – Shell and Gulf being responsible for 40 and 7 per cent respectively. In addition, Jersey and Shell became the dominant oil companies in Venezuela and the operations of the other members of the Seven Sisters who eventually established themselves in the country (the Texas Company, Standard of California, and Socony-Vacuum[13]), were never able to emulate the scope or success of either.

Until the 1950s, only companies belonging to the Seven Sisters produced oil in Venezuela. After the concession acreage auctions held under the auspices of the Pérez Jiménez government in 1956–7, however, the range of concessionaires widened noticeably. In these bidding rounds, Jersey, Gulf and Shell were able to obtain 'only' 43 per cent of the total area open to bids, because of the high premiums per hectare paid by other companies, mainly American independents eager to establish a foothold in Venezuela. The number of foreign concessionaires was destined never to expand again after the 1956–7 auction rounds. Shortly after the auction, democratic forces ousted Pérez Jiménez from office, and Rómulo Betancourt, of Acción Democrática, assumed the presidency of the republic. Among the first actions of the new government was the reinstatement of the ban on any new concession grants. The Betancourt government also established a national oil company, Corporación Venezolana del Petróleo (CVP). CVP was granted acreage relinquished by private oil companies to explore and develop, as well as a small refinery at Morón. In 1963, the government decreed that foreign oil companies were to relinquish the control of one-third of the Venezuelan domestic market to CVP by 1968. Nevertheless, CVP never became a major oil company, able to provide the Venezuelan government with a sound foundation on which to build a national oil industry. In 1975, at the time of the nationalization, CVP was responsible for less than 4 per cent of the country's oil production, and yet it accounted for 13

Table 2.1: Oil Companies Holding Concessions in Venezuela. 1974.

Company	*Parent organization*	*Concession area (acres)*	*Percentage of production (1973)*
Amoco Venezuela	Amoco	13,591	0.96
Charter Venezuela	Charter Venezuela	17,297	0.33
Chevron Venezuela	Chevron	192,399	1.34
Continental Venezuela	Conoco	1,969	0.10
Coro	Texaco	138,042	–
Creole	Exxon	1,433,827	44.43
International Petroleum	Exxon	15,155	3.01
Mene Grande	Gulf, Exxon, Shell	1,467,151	6.02
Mobil de Venezuela	Mobil	355,958	3.16
Phillips	Phillips	97,476	0.90
Sinclair Venezuela	Arco	96,571	–
Talon	Kirby	148,673	0.10
Texas	Texaco	271,579	2.09
Texaco Maracaibo	Texaco	7,776	1.63
Venezuela Atlantic	Arco	47,782	2.38
Venezuela Sun	Sun	49,420	2.36
Caracas	Ultramar	68,476	0.28
CVP	Venezuelan government	2,914,144	2.20
Mito Juan	Mito Juan	85,458	0.09
Petrolera Las Mercedes	Texaco-Ultramar	220,436	–
Shell de Venezuela	Royal Dutch/Shell	730,625	26.66
Others*	–	–	1.96

*Ashland, Kerr McGee, Monsanto, Murphy, Pacific Petroleum, Petrofina, Tenneco, Union Oil

Source: Tugwell, 1975, p. 10

per cent of the workforce of the industry. (Table 2.1 shows the companies producing oil in Venezuela just before the nationalization.)

After Venezuela nationalized all concessions, it established PdVSA as the new national oil company. PdVSA then signed a series of technical service agreements with its most important ex-concessionaires. In return for a fee for each barrel of oil produced and, where applicable, for each barrel of crude refined, by their former holdings, these companies would provide PdVSA with specialized services for production, refining, training and acquisitions. These agreements also contemplated the 'loan' of foreign personnel to the new Venezuelan operating companies. The increasing technical knowledge of PdVSA gradually made these agreements redundant, and they have been allowed to expire. However, if PdVSA's plans for reactivating marginal oilfields in Venezuela go ahead (see Chapter 3), it could refer to the precedent set by these service contracts when negotiating terms with its new contractors, whoever they might be.

2.4 PdVSA: Corporate Structure

Article 6 of the nationalization law – also known as the Law Reserving the Production and Marketing of Hydrocarbons to the State – specified that the Venezuelan executive would be able to 'create the enterprises necessary for the regular and efficient development of [all activities relative to the exploration, exploitation, refining and marketing of hydrocarbons]', assigning to one of these enterprises 'the duties of co-ordination, supervision and control of the activities to be undertaken by the rest of the companies' (CEPET, 1989, v. II, p. 423). On 30 August, 1975, executive decree 1123, based on this article, created a new holding company, Petróleos de Venezuela S.A. to take over these co-ordination, supervision and control duties. When the nationalization law became effective, on 1 January, 1976, the 22 concessionaires with operations in Venezuela ceded their places to 13 nationalized operating companies (plus CVP).

The reason for this awkward arrangement – at first sight, it would have been easier to amalgamate the smaller companies with the three largest, Lagoven (successor to Creole), Maraven (successor to Shell) and Meneven (successor to Mene Grande), which accounted for 82 per cent of Venezuelan production – was not difficult to fathom: all the operating companies were based as closely as possible on established divisions of ownership,[14] something which preserved the important differences in management style, managerial talent and technical expertise which existed between the operating companies. Thus, the fact that the companies continued working under existing management structures permitted PdVSA to avoid the productive disruptions which would have arisen if a major amalgamation of the smaller companies had taken place immediately after the nationalization. This notwithstanding, it had been clear from the very onset of the nationalization, that the number of operating companies would have to be drastically reduced in the future, if PdVSA's operations were to become reasonably efficient, because 'the fourteen operating companies . . . were too different in terms of size, technical knowhow, and strength of labour force [for all of them] to be able to survive' (Coronel, 1983, p. 109). Thus, in July 1976, the five largest operating companies – Lagoven, Maraven, Meneven, Llanoven and CVP – produced a document which would serve as the blueprint for the administrative rationalization of the industry. This document divided the operating companies into five groups: companies with very large organizations (Lagoven, Maraven, Meneven); medium-sized companies, needing substantial reinforcing to be able to compete with their three big brothers (Llanoven, Deltaven, Palmaven); small companies which would have to be integrated into larger organizations in the short term (Boscanven, Roqueven, Bariven, Amoven); marginal companies, which would have to be integrated into larger companies at once (Vistaven, Taloven, Guariven); and finally, CVP, the original state oil company (Coronel, 1983, p. 111). The document stressed that the first group possessed 85 per cent of the proved oil reserves, 82 per cent of the production capacity, 84 per cent of the refining capacity and 72 per cent of the labour force. Taking this into consideration, the presidents of the biggest operating companies recommended that the number of operating companies be gradually reduced, to a maximum of four. To achieve this,

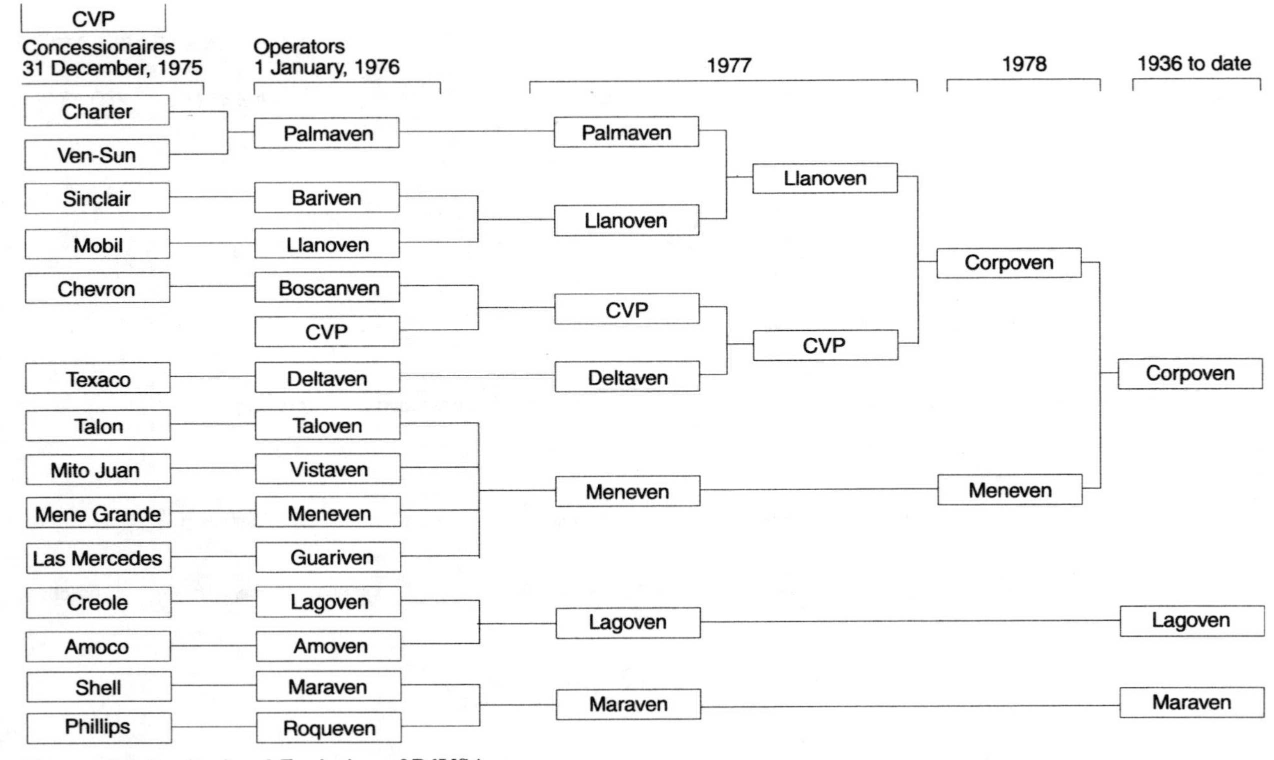

Figure 2.2: Institutional Evolution of PdVSA.

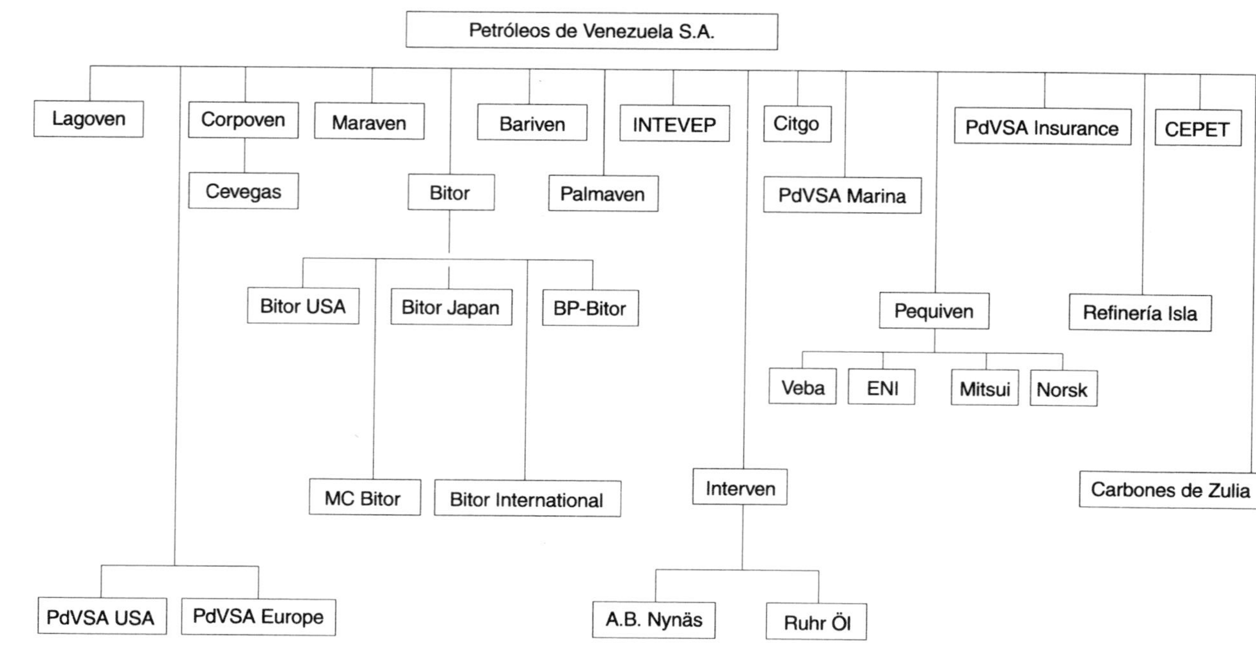

Figure 2.3: PdVSA Corporate Structure

the document proposed a two-stage plan. First, the larger companies would begin to serve as advisory agents for smaller companies and after a suitable period of time they would absorb them into their organization. In December 1978, the smaller producing companies were merged into PdVSA's major operating subsidiaries. Three companies were taken over by Meneven, one by Lagoven and one by Maraven, and a new company, called Corpoven, merged the assets of the remaining five companies with those of CVP (see Figure 2.2). This left the number of operating subsidiaries at the recommended level of four, but a further reorganization of operations was undertaken in June of 1986, when Meneven was abolished, and its interests taken over by Corpoven.

The completion of the rationalization process left PdVSA with a unique corporate structure: a holding company with, among other things, three completely vertically integrated oil companies as subsidiaries. The dynamics of this arrangement are hard to understand, even for the Venezuelans themselves:

> To some, the three [companies] are brothers, each seeking to excel against the world in the interests of all. To others, they are again three brothers, yet this time fighting against each other to be top dog. To outsiders, the ambiguity seems to reach its oddest form, when three sets of logos dominate three different types of gasoline service sections. Here a ferocious marketing fight seems to be going on between three identical products to send money to exactly the same ultimate destination (*EE*, December 1991, p. 15).

To keep this system functioning smoothly, PdVSA has always taken great care to foster competition between its different operating companies. At times, it has intervened in the management of affairs to ensure that none of them achieves a preferential position *vis-à-vis* the others. For instance, when Corpoven took over the personnel operations of Meneven, the company's refining and production activities to the west and north of the Andes were given to Maraven, and its exploratory acreage in Lake Maracaibo to Lagoven. Also, even though it was Lagoven's R&D efforts that led to the development of the Orimulsion boiler fuel (see Chapter 4), PdVSA has arranged its Orimulsion operations in such a way that

Table 2.2: PdVSA. Oil Production by Subsidiary, and Type of Crude. 1988–90. Billions of Barrels.

		0–21.9° API	
	1988	1989	1990
Corpoven	29.323	21.379	35.285
Lagoven	50.792	43.006	55.487
Maraven	73.597	51.544	75.124
		22–29.9° API	
	1988	1989	1990
Corpoven	67.144	86.384	104.030
Lagoven	197.040	198.222	198.694
Maraven	13.060	15.369	17.839
		30+ ° API	
	1988	1989	1990
Corpoven	44.389	49.234	54.317
Lagoven	50.375	53.538	59.407
Maraven	170.892	177.655	178.731

Sources: MEM, 1990, p.58.

Maraven and Meneven will not be left out of this potentially lucrative enterprise. Table 2.2 shows the production of crude oil by subsidiary in the last few years.

Pequiven is arguably the most important PdVSA subsidiary after Lagoven, Maraven and Corpoven. It was created in 1977, by means of a law which converted the Instituto Venezolano de Petroquímica (Venezuelan Petrochemical Institute) into a company whose shares would be owned by PdVSA, who was given the extremely unappetizing task of absorbing IVP's huge debts, and turning around the fortunes of the industry. In this, it has been quite successful: in 1977, the Venezuelan petrochemical industry was producing 800,000 tonnes a year of products on a 2.1 million tonne (mt) production capacity; in 1990, total production reached a level of 2.2 mt, of which 0.5 mt were exported.[15]

As Figure 2.3 shows, PdVSA has a number of other operating subsidiaries. Some of these, like Bariven and Palmaven, were part of the original panoply of its subsidiary oil-producing companies. When these companies were rendered inactive by their absorption into bigger companies, PdVSA decided to maintain their legal status, which made it possible for them to be reorganized and given other functions (CEPET, 1989, v. 11, p. 429).[16] Other non-oil subsidiaries include Bitor (created to market the Orimulsion boiler fuel) and Carbones de Zulia S.A., whose purpose is the development of the Guasare carbon mine project. PdVSA's two newest subsidiaries are PdVSA Marina and PdVSA Insurance.[17]

Other oil-related subsidiaries of PdVSA are Refinería Isla S.A. – created to operate the refining complex and maritime terminal which PdVSA leases in Curaçao – and the PDV-America and PDV-Europe companies, domiciled in Caracas, but with offices in New York and Houston, and London, respectively. Their mission is to provide PdVSA and its other subsidiaries with timely and relevant market intelligence. To improve the Venezuelan oil industry's technological autonomy, PdVSA took over the government agency INVEPET, and gave it full subsidiary company status, under the name INTEVEP. Its mission was to establish laboratories and test facilities, in order to enhance the PdVSA's R&D capabilities. To complement INTEVEP's activities, PdVSA and its subsidiaries constituted CEPET, a civil association whose purpose is the formation, training and professional development of the personnel of the petroleum and petrochemical industries.

2.5 PdVSA and the Ministry for Energy and Mines: the Triumph of Politics over Profits

After the nationalization of the oil industry in 1976, the basic organizational scheme of the Venezuelan oil sector was defined as being composed of three administrative levels, 'with clearly defined objectives, structures and systems' (CEPET, 1989, v. II, p. 204). The ministry of energy and mines was made the 'maximum organ for the formulation of policy' (ibid.) The second administrative level is occupied by PdVSA, 'which functions as the mother company of the nationalized petroleum industry and is charged with strategic planning, coordination, supervision and control of its subsidiaries' which, in turn, constitute the third and last, level, and are in charge of the execution of the plans and programmes of the oil industry (ibid.).

The division of labour inherent in this bureaucratic structure ideally would be quite straightforward. The ministry would be responsible for setting down broad policy guidelines; PdVSA would formulate specific programmes to achieve the goals contained in the ministry's policy statements, and oversee their correct implementation by its subsidiaries. Finally, the operating companies would be in charge of the execution of these programmes. However, there are some factors that complicate this neat arrangement. On one hand, the exact reach of PdVSA's and the ministry's jurisdictions is not at all clear (as can be discerned from the above paragraph), and has traditionally been made even more nebulous by the fact that decrees antedating the nationalization, which gave the ministry a very high level of control over the industry, were never explicitly revoked. Thus, 'ministry staff argued from the very first day of nationalization that decree 832[18] was still in force and that, therefore, all nationalized industry activities had to have their previous approval', while in the opinion of Andrés Aguilar, legal adviser to PdVSA, 'decree 832 had been automatically rendered invalid' by the nationalization (Coronel, 1983, p. 106). This lack of clear delimitations of administrative responsibilities has led to conflicts between the minister of energy and the president of PdVSA, especially when these two officials do not see eye to eye.[19]

On the other hand, the fact that, from its very inception, PdVSA was given a mandate to operate as a private, profit-oriented, company,[20] ostensibly immune from any political interference,[21] is an element which can also cause problems in the relationship between the ministry and the company. As many parts of this study will show, it is not difficult for situations to arise which lead to PdVSA and the government having incompatible political and entrepreneurial goals. This is because a state company like PdVSA, unlike a private company (which can be said to have a single, unchanging primary objective, the pursuit of profit maximization), has to contend with ever-changing objectives, which are not always very well defined, and often contradict each other. The tacit rules governing its conduct fluctuate according to the vagaries of public opinion or bureaucratic whims, sometimes giving precedence to profit maximization, other times to the pursuit of social objectives. For a state oil company, in particular, these swings mean that it has to chart a course of action between 'a political level that is essentially concerned with the government's ability to survive and maintain its freedom of action amid conflicting pressures' and 'an oil level, which is concerned with practical solutions to problems directly related to the oil industry' (Noreng, 1980, p. 111).

In sum, contradictions in the operations of a state oil company are bound to develop sooner or later. However, when they do, it is very important to see which set of values is dominant, the one associated with profits or the one associated with politics. In the case of PdVSA, many people tend to assume that, traditionally, it has maintained its operations reasonably free from government interference, and kept at bay the attempts of Venezuelan politicians to control it more directly.[22] Unfortunately, this is not the case. Granted, the company has foiled many attempts to place it under the *de jure* control of the energy ministry,[23] but, it has not been able to prevent a *de facto* control of its operations. The prime example of this, undoubtedly, is its failure to prevent the bankrupt government of Luis Herrera Campíns from confiscating its $18 billion worth of investment[24] funds in September 1982, after having fought off initiatives that would have placed it under direct government control. This failure was very significant, because it effectively ended PdVSA's status as a self-financing company. Moreover, this pattern whereby PdVSA wins the battles but in the end loses the war for control was repeated in August 1991. Citing orders from the president of the republic, energy minister Celestino Armas sent a directive to PdVSA president Andrés Sosa Pietri which contained measures that radically altered the terms of interaction between the ministry and PdVSA. Armas' directive made PdVSA (and all its domestic and overseas affiliates) unable to create new affiliates or change the bylaws of its existing affiliates without prior approval from the ministry. Likewise, ministerial approval was needed before it could embark on any acquisition of new assets through mergers, liquidations or restructuring, contract bank loans in Venezuela or abroad, issue bonds or any other financial instrument, purchase productive assets or contract project financing. Finally, PdVSA was rendered unable to incur any 'costs and expenses' not anticipated in its government-approved budget, to make any decision regarding salaries and related benefits, pensions and retirement plans for the industry's managers, or make board appointments to its related companies (*PON*, 21 August, 1991, p. 4).

Quite rightly, Armas' memorandum was seen as a massive infringement of PdVSA's entrepreneurial autonomy, and the whole board of the company threatened to resign in protest. This made Carlos Andrés Pérez reconsider his options,[25] and a few days after the storm had broken, he issued a directive aimed at preventing further conflicts arising between the ministry and PdVSA. The new directive allegedly restored PdVSA's administrative autonomy, because, even though it maintained all the restrictions contained in Armas' original memorandum, it dropped the requirement for prior approval by the ministry (PdVSA was instead required to 'inform' the ministry before doing any of the restricted things). The directive also created a special commission composed of the oil minister, the president of PdVSA, the planning minister and the head of the Central Bank of Venezuela, 'to insure harmony in the execution of oil investments and the government's macro-economic and financial programming' (*PON*, 23 August, 1991, p. 1). However, PdVSA's victory (such as it was) was Pyrrhic, because a few months later, the government slashed its 1992 budget by more than a billion dollars, and froze some important areas of the company's 1991–6 expansion plans.[26] This decision, as we shall see, puts most of the industry's expansion plans very much in question, and it could even compromise the long-term perspectives of Venezuela as an important oil exporter.

The issue of the relationship between PdVSA and the central government in Venezuela has been very aptly encapsulated in a metaphor by Gustavo Coronel, one time vice-president of Meneven. Coronel compared the company to the New York Yankees baseball team: 'the Yankees', said he, 'are the richest team with the best players, [but they] have the greatest organizational problems because their owner wants to run the team from the stands' (Coronel, 1983, p. 213).

2.6 Venezuela and OPEC

As we have seen, the foundation of OPEC was the result of the travails of two men: Juan Pablo Pérez Alfonzo, Venezuelan minister of energy and hydrocarbons, and Abdullah Tariki, Saudi oil minister. In 1960, they forged an alliance of oil producers (initially composed of their two countries and Iran, Iraq and Kuwait) which, a decade later, was perceived as powerful enough to strangle the economies of the first world (as Henry Kissinger implied when he accused the Arab nations in OPEC of waging economic war against the West).

Pérez Alfonzo's idea for a cartel-like arrangement of oil producers originally stemmed from his keen dislike of the profligacy and waste, a situation which he perceived as endemic in the oil market:

> To sell oil cheaply, he argued, was bad for consumers, as the result would be the premature exhaustion of a nonrenewable resource and the discouragement of new development. For the producing countries, oil was a national heritage, the benefits of which belonged to future generations as well as the present. Neither the resource nor the wealth that flowed from it should be wasted . . . Human nature should not be allowed to squander the potential of this precious resource (Yergin, 1991, p. 512).

However, his enthusiasm for OPEC's adoption of pro-rationing methods like those of the Texas Railroad Commission was not just due to his belief in the validity of Hotelling's economics. He also thought that a pro-rationing system would allow Venezuela to defend its market share effectively, by preventing her main markets from being swamped by low-cost oil from the Middle East.

As an organization dedicated to the conservation of oil, OPEC was a source of great disappointment for Pérez Alfonzo.[27] After him, there have been other Venezuelan politicians who have also grown disenchanted with the organization, and Venezuela's apparent lack of leverage within it. Thus, as a result of different disputes within the organization (but especially the infamous 'condensate dispute'[28] involving Venezuela, Kuwait and Saudi Arabia), a perception of OPEC as an Arab-dominated lobby, whose decisions are not always in Venezuela's best interests, has taken root in the country (and, in particular, in PdVSA). Of late, especially during the term in office of PdVSA ex-president Andrés Sosa Pietri, there have been abundant calls for Venezuela to abandon OPEC.

The argument usually wielded to justify this course of action is that the OPEC quota system unnecessarily shackles Venezuela's oil production. For instance, when he announced his resignation as president of PdVSA, Sosa Pietri remarked: 'I have been in absolute and complete disagreement with the [OPEC] limits on the nation's oil production, and I consider them absolutely contrary and contradictory to [PdVSA's interests]' (*EC*, 3 April, 1992). His arguments, however, do not stand up to close scrutiny.

In his study about the OPEC quota system, Robert Mabro has shown that one of the effects of the implicit principles underlying the allocation of quotas would be to diminish the weight of countries, like Venezuela, which do not have a sizeable excess production capacity (or, other things being equal, large reserves[29]), because 'the [crucial] problem of how to share incremental demand soon becomes the *exclusive* problem of five Gulf countries . . . Iran, Iraq, Kuwait, Saudi Arabia and UAE' (Mabro, 1989, p. 30). In other words, Venezuela's diminution of power within the organization reflects the realities of her oil sector, rather than an Arab-led conspiracy to undermine it. Furthermore, it is also extremely unlikely that, in the absence of OPEC quotas, Venezuela would be able to produce crude to her heart's content. This is because any incremental capacity it brought on stream would produce very heavy and sour crudes, and the international market's absorption capacity for this type of crude is limited. This last aspect was proved to be true in the aftermath of the Iraqi invasion of Kuwait. Initially, amidst massive market uncertainty, oil consumers were glad of any incremental supplies of crude which they could get their hands on, even if they came from Venezuela. However, as soon as the situation stabilized somewhat, and the crisis apparently reached a stalemate, PdVSA started to experience difficulties in marketing her incremental crude supplies.

Bearing all this in mind, we must then ask, what are the probabilities that Venezuela will abandon OPEC? Back in 1989, Mabro explained that, since the role which OPEC played in the world market depended basically on the behaviour of the Persian Gulf countries with large oil reserves, it was appropriate to ask 'whether OPEC will become in the near future a much smaller organization consisting of the five large reserve Gulf countries [Iran, Iraq, Kuwait, Saudi Arabia and the UAE] and one or two non-Gulf countries' (Mabro, 1989, p. 56). However, when he examined the cases of all the countries with little or no apparent economic incentives to stay within the fold of the organization (read Venezuela), he concluded that 'Venezuela derives political kudos from its historical role in the formation of OPEC.' Furthermore, he raised a question which some of the Venezuelan politicians who have recommended leaving OPEC would find uncomfortable to answer from a political viewpoint: 'How can Venezuela abandon its begotten child?' (ibid., p. 57).

Mabro's conclusions are as valid now as they were then. Even if one accepts that, production-wise, Venezuela's actual role in OPEC is limited to being one of the countries that 'hold a bargaining card, since they can expand oil output by amounts that are small compared with surplus capacity in the Gulf, yet sufficiently large to cause a temporary price disruption' (ibid., p. 56), it is accurate to say that Venezuela still reaps substantial political benefits from its continued OPEC membership. In my opinion, for instance, Carlos Andrés Pérez's role as the co-organizer of the 1991 Paris meeting between the main oil producers and consumers of the world would have enjoyed far less credibility had his country not been perceived as a principal member of the most important association of oil producers. Thus, notwithstanding PdVSA's dislike of OPEC, it is unlikely that, for as long as the ministry of energy and mines and the Venezuelan executive have any say in the matter, the country will be leaving OPEC.

Finally, it should also be made clear that the frustration and animosity which some Venezuelan politicians (or PdVSA bureaucrats) feel towards OPEC, and their impression that it is a forum in which Venezuela's voice is consistently ignored, has as much to do with ignorance about Middle Eastern affairs as it does with the geographical and geological factors that make the 'Gulf bloc' the real centre of power in the organization. It is clear that, after 32 years of meeting with their Gulf counterparts, there is still not enough understanding of each other's positions.

Notes

1. Even though oil was commonly known as *stercum demonis*, or devil's excrement. Some centuries later, a former Venezuelan oil minister would bring this ancient usage, in a much publicized harangue against black gold.
2. Throughout the revolution, the oil companies paid off different revolutionary warlords, who in turn protected the companies' properties.
3. For a full description of the dangers faced by early oil prospectors in Venezuela – ranging from extremely vicious parasites to Indians with deadly poisoned arrows – see Yergin, 1991. pp. 233–7.
4. In 1948, just before Pérez Jiménez's coup, the oil companies – in response to the nationalistic stance of Acción Democrática – laid off 18,000 petroleum workers (one-third of the total), cut production by 15 per cent (200,000 b/d), cut off all investment in Venezuela and, in general, shifted their sights to the Middle East. The dictator's policy of appeasement brought them back in droves, with such success that, in the years between 1948 and 1957 (when Pérez Jiménez was toppled), production doubled, going from 1,339 mb/b to 2,779 mb/d (Lieuwen, 1985, p. 204).
5. The Arab nations and Iran were not convinced by Pérez Alfonzo's arguments about the necessity of conservation. As Evans says, 'only Venezuela, OPEC's largest producer throughout the 1960s, was strongly in favour of a collective agreement to control the rate of growth of member countries' output with a view to supporting the level of oil prices in the world market on the basis of prorationing' (1991, p. 76).
6. Before this law, concessions had been granted under general mining laws which defined the government's royalty as a fixed sum per tonne of production.
7. For instance, under the 1922 amendments to the law, concession periods were extended to 40 years, royalties set at a lower level of 10 per cent (and 7.5 per cent or underwater acreage in Lake Maracaibo), and customs exemptions granted for industry-related imports.
8. The phrase was coined by Gumersindo Torres, Gómez's minister or development on two separate occasions. Torres thought, quite rightly that in nowwhere other than Venezuela were the oil companies allowed to take so much and give so little in return' (Lieuwen, 1985, p. 197). Because of his ultimately unsuccessful attempts to defend the country's interests, Torres became a folk hero of sorts in Venezuelan oil circles (see Betancourt, 1978, pp. 20–22; and Martínez, 1980, *passim*).
9. New concessions covering an area of 6,500 sq. km. were granted in 1944–5. Some of the acreage by companies with no existing Venezuelan interests (Mobil, Texaco).
10. In 1940, Shell and Jersey had a refining capacity of approximately 495,000 b/d in the Netherlands Antilles, and only 45,000 b/d in Venezuela. Overall, about 95 per cent of the country's output was refined abroad at this time.
11. The same system is used to tax PdVSA today.
12. The following 11 companies accounted for 94.1 per cent of total compensation payable: Creole, 45.9 per cent; Shell de Venezuela, 24.13 per cent;

Mene Grande, 6.68 per cent; International Petroleum, 3.55 per cent; Texaco Maracaibo, 2.98 per cent; Venezuela Sun Oil, 2.64 per cent; Mobil de Venezuela, 2.22 per cent: Phillips Petroleum, 1.92 per cent; Venezuelan Atlantic Refining Company (ARCO), 1.72 per cent; Amoco Venezuela, 1.23 per cent; and Chevron of Venezuela, 1.16 per cent.

13. British Petroleum never operated in Venezuela because of a Venezuelan law prohibiting companies owned by foreign governments from being given any concessions.
14. Nine concessionaires, however, were not incorporated as nationalized operating companies. These were: Caracas Petroleum (Deltaven); Conoco Venezuela (Maraven); Coro Petroleum (Deltaven); Charter Venezuela (Palmaven); Eastern Venezuela Gas Transport Company (CVP); International Petroleum (Meneven); Texaco Maracaibo (Deltaven); Venezuelan Atlantic Refining (Bariven and CVP); and Venezuelan Gulf Refining Company (Meneven). The names in parentheses indicate the nationalized companies that received the assets and personnel of these firms (CEPET, 1989, v. II, p. 428).
15. Everything seems to indicate that, due to the severe budgetary constraints imposed on PdVSA by the government in 1991, Pequiven will be transformed into a joint stock company. Shares in this company will be offered in the most important stock exchanges in the world.
16. Bariven is in charge of PdVSA's purchases and acquisitions abroad, while Palaven has been entrusted with the production of fertilizers. Converted into a Pequiven (q.v.) subsidiary after 1978, Palmaven now enjoys full PdVSA subsidiary status, ostensibly in order to permit 'bigger and better cooperation with [Venezuela's] agricultural sector by the oil industry' (CEPET, p. 207). Another of the original oil subsidiaries which was assigned new functions is Amoven. Renamed Interven, its original mission was the co-ordination and direction of all the activities and investments undertaken by PdVSA abroad. Recently, however, PdVSA's overseas operations have been restructured. Its American operations have been merged and put under the aegis of the Citgo Petroleum Corporation, its most important American subsidiary, thus reducing the scope of Interven to operations in Europe.
17. The former has the responsibility of operating the extensive fleet of ships owned by the company, as well as controlling chartering arrangements and the like; the latter of course takes care of activities in the field of insurance.
18. Introduced in the early 1970s, this decree required concessionaires to submit a copy of their operational programme for the year in advance to the ministry for approval and, if necessary, modification.
19. As was the case with General Rafael Alfonzo Ravard and Humberto Calderón Berti in the early 1980s, or Celestino Armas and Andrés Sosa Pietri in the early 1990s.
20. Thus, the company was granted control over its financial resources (its statute allows it to keep 10 per cent of its income from exports for reinvestment) and obliged to present full audited accounts, due to its officially

'private' nature. Article 8 of the nationalization law decreed that the directors, managers, and workers of the nationalized oil industry are not considered public employees.

21. This *laissez faire* approach was not as radical as that of the British government regarding British Petroleum, however. The company was ultimately responsible to the president of the republic and the minister of energy and mines sat as president of PdVSA's one member stockholders' annual general meeting. The Venezuelan congress can also interfere in PdVSA's operations, since it can call hearings into the company's operations and finances, and its approval is required for any operation calling for foreign participation in the oil sector.
22. See for instance the editorial 'A Policital Setback in Venezuela' in the 2 September, 1991 issue of *O&GJ*.
23. Many Venezuelan politicians have looked towards PEMEX as an institutional (although never an operational) model for PdVSA. Andrés Sánchez Bueno, justice minister during Pérez's first term in office, said: 'The Mexican Congress approves the budget of PEMEX . . . while . . . in Venezuela . . . we have not been able to ensure that state-owned enterpises submit their investment proposals and budgets to Congress for approval. We should pass a law in this respect' (Coronel, 1983, p. 192). Likewise, in 1979, young congressman Cestino Armas stated that the oil industry enjoyed excessive freedom, and measures should be enacted to curb it (ibid., p. 183).
24. These funds had been accumulating since 1976. Herrera used them for, among other things, bailing out the corrupt and scandal-laced Workers' Bank (see Coronel, 1983, chapter 12).
25. Celestino Armas was accused of being the prime mover behind the initiative, out of his personal dislike for Sosa Pietri, but it is clear that he could not have taken the decision without the explicit consent of the president.
26. It is interesting to note that when Herrera Campíns seized PdVSA's investment fund, Carlos Andrés Pérez decried the move a jeopardizing the future of the Venezuelan oil industry. Thus, his attitude towards measures which could turn out to be just as damaging to the industry, can only be called quite contradictory.
27. Pérez Alfonzo said in 1979, 'OPEC, as an ecological group, has really disappeared' (Yergin, 1991, p. 525).
28. See Chapter 3, note 12.
29. This point would also be valid for Venezuela, given the particular composition of her crude reserves.

3 THE VENEZUELAN UPSTREAM SECTOR

3.1 Producing Areas and Oil Fields

On 14 December, 1922 Venezuela became one of the world's most important oil provinces when Royal Dutch/Shell's Los Barrosos-2 well (located in the La Rosa area of the Bolívar Coastal Field) blew out. The well gushed without control for nine days, at a rate estimated at 100,000 b/d, until it finally sanded itself up and stopped. This spectacular incident, coming as it did on the heels of the Mene Grande and Las Cruces discoveries (which took place in April 1914 and August 1916, respectively), transformed Venezuela into an oil powerhouse. Since then, several hundred more oil accumulations have been found, with at least 45 among them being US-type 'giants' (that is, fields containing at least 100 mb of oil).[1] Eleven of these fields have cumulatively produced more than 500 mb since their discovery date (up to 1990; see Table 3.1). The cumulative oil produced in Venezuela, up to 31 December, 1990, has been 44.4 billion barrels (*O&GJ*, 30 December, 1991, p. 87).

Venezuela's oilfields are found in four major sedimentary basins: Maracaibo, Falcón, Barinas-Apure and Oriental, which together cover about 43 per cent of the total area of the country. The Maracaibo basin is by far the most important of these: as far as oil provinces go, it is surpassed in size only by the Arabian-Persian Gulf province. Its awesome dimensions defy the imagination. Approximately 37 billion barrels of crude oil have been extracted from it, and still the 1990 crude production from the basin – 1,407 mb/d – accounted for 68 per cent of the Venezuelan total.

The Maracaibo basin has an extension of 50,000 sq. km., and is surrounded on three sides by mountains: the Andes to the south, the Perijá and Montes de Oca ranges to the west and the Trujillo range to the east. To the north, it is separated from the continental shelf of the Gulf of Venezuela by the Oca fault. The central part of the basin is occupied by Lake Maracaibo, a relatively shallow body of water connected to the

Table 3.1: Main Oil-producing Fields in Venezuela

Name and date of discovery	*Cumulative production to 31 Dec. 1990 (Billion Barrels)*	*1990 average production (b/d)*
Lagunillas, 1926*	11,086,650,262	361,093
Bachaquero, 1930*	6,631,194,318	245,929
Tía Juana, 1928*	4,110,081,045	141,300
Lama, 1957*	7,679,286,645	135,629
Cabimas, 1917*	757,784,958	46,880
La Paz, 1925	865,729,055	9,252
Quiriquire, 1928	761,694,081	1,075
Centro, 1959*	991,944,863	148,065
Boscán, 1946	803,178,000	28,851
Mene Grande, 1914	637,755,243	340
Lamar, 1958*	1,192,397,969	81,396
Ceuta, 1985*	500,635,876	63,957
Mara, 1945	417,706,447	5,804
El Furrial, 1986	129,833,355	132,946
Musipán, 1985	116,836,387	49,180
Pirital, 1958	27,535,135	75,439
Guafita, 1984	66,703,939	45,980
Carito, 1988	33,390,636	53,826
Motatán, 1952	118,241,052	48,614
Urdaneta, 1970	168,221,172	42,399

*Bolívar Coastal Field
Source: *O&GJ*, 30 December, 1991, pp. 86–7.

Caribbean sea by a strait. In this area, says Tiratsoo, 'sediments accumulated . . . in an oxygen-deficient sinking basin or embayment frequently cut off from the sea', whose geological conditions, like those of the Gulf of Mexico or the Caspian Sea, 'were nearly ideal for petroleum generation' (1984, p. 366).

Lake Maracaibo divides the basin into three zones: the Occidental Coast, the lake area proper and the Oriental Coast (also known as the Bolívar Coast). In the eastern seaboard of the lake, most of the crude deposits are located in shallow Miocene formations, and contain heavy and medium crudes. The deposits in the central lake area are basically geological extensions of the eastern coastal fields, and they can be found at between 1,200 and 5,000 metres depth, with reserves of medium and light crude oils. These deposits go back to the Cretaceous, the Eocene and the Miocene, while west coast formations are mainly from the Cretaceous, and contain light oils (CEPET, 1989, v. 1, p. 205).[2] In other words, crudes in the basin show a tendency to segregate themselves vertically according to their age and gravity. Thus, lighter crudes and condensates are generally found in older and deeper formations, while the heavier crudes lie in more recent, shallower, ones (CEPET, 1989, v. 1, p. 204).

The fields in the eastern coast of the lake cover an area of roughly 2,000 sq. km., nearly 60 per cent of which lies offshore. The word 'fields', however, is somewhat of a misnomer. In reality, the oil deposits in the area are all part of one single mammoth accumulation, the Costanero Bolívar field (known in English as the Bolívar Coastal Field or BCF), which has been divided into separate units purely on statistical and administrative grounds. BCF extends for more than 100 km. along the shore of the lake and it includes at least 25 giant oilfields (US-type), some of which can be counted among the most prolific in the history of oil (Lagunillas, Bachaquero, Tía Juana).

For the better part of 40 years, production in the onshore part of BCF was carried out relying on the natural flow of the fields and using only mechanical pumping. In the middle of the 1960s, drops in reservoir pressure prompted the introduction of steam soaking exploitation techniques in the most important fields. Gradually, steam soaking has come to be considered a conventional production method, indispensable in the task of extracting viscous crudes from shallow formations (CEPET, 1989, v. 1, p. 208). Offshore, the introduction of thermal enhanced recovery techniques has been deferred by the less viscous nature of the crude subject to extraction. However, it has been necessary to resort to other, less expensive, secondary recovery techniques. Gas injection began to be used in the Lagunillas field in 1954; in 1967, water injection was first applied to the Bachaquero field. The number of wells in BCF producing by means of these methods has grown steadily ever since (see the maps in CEPET, 1989, v. 1, pp. 197–8).

The west coast of the lake has traditionally been known for its light and medium crude fields (La Paz and Mara). Heavy crude, however, is present in significant quantities in the Boscán field. The oil in Boscán is found at an average depth of 2,500 metres. This confers a productive advantage because the higher temperatures registered at such depths diminish the viscosity (resistance to flow) of the crudes. However, this same depth makes the application of traditional production methods for viscous crudes quite difficult (CEPET, 1989, v. 1, p. 215). The extreme lenticularity of the sands present in this deposit also has an adverse effect on crude extraction.

The Maracaibo basin extends right up to Venezuela's border with Colombia. The Tarra group of fields (Las Cruces, Los Manueles, Tarra Oeste) is located in this area. To the north and north-west of the Tarra group, lies yet another group of fields (Río de Oro, Rosario, Alturitas), which has not been extensively exploited due to lack of pipeline outlets (Tiratsoo, 1984, p. 367).

The Oriental basin is the second largest oil province in Venezuela. Although it had been well known from the early part of the twentieth century for the presence of oil seepages,[3] the first commercial discoveries of oil in the Oriental basin were not made until 1937. The basin is divided into two sub-basins: Guárico and Maturín. The Guárico sub-basin is located in the federal state of the same name. Its commercial production is concentrated in the area of Las Mercedes, to the south of the basin. Another group of fields, collectively known as the Valle de Pascua group, is located to the east of Las Mercedes. None of the Guárico fields, however, is currently of any importance in terms of Venezuelan crude oil production. Quite the opposite is true with the fields of the Maturín sub-basin (which lies in the federal states of Anzoátegui and Monagas): their production in 1990 was equivalent to 27 per cent of the Venezuelan total. In the centre of the basin, a great many oilfields have been discovered, and are known collectively as the Oficina group.[4] To the north and northwest of Oficina lie the fields of the Anaco group (Santa Ana, San Joaquín, Santa Rosa and El Toco). To the south of the basin, there exist a number of producible heavy and extra-heavy oilfields (Pilón, Jobo, Morichal, Temblador), which grade into the hydrocarbon-rich Orinoco Oil Belt. Finally, on a trend running northeast from the Oficina fields one can find the Jusepín group, whose oilfields are extremely important because they produce light and medium crudes. In fact, the light and medium crude oilfields discovered since 1986 in the Maturín sub-basin (El Furrial, Musipán) have made an important contribution to stemming the precipitous decline in Venezuelan non-heavy oil production (see section 3.4), and play a critical role in PdVSA's future E&P plans.

The Falcón basin is dwarfed in size by the other Venezuelan oil provinces. Its production has always been quite small, and marginal in character. The deposits in the basin consist of light, highly paraffinic crude, with low sulphur and metal contents. Most of the oil wells in the basin are shut-in, although plans do exist which consider bringing them back on-stream (see section 3.6). The most important oilfield yet found in Falcón is Cumarebo, which has produced a cumulative 50 mb of oil (and was shut-in in 1990). The Barinas-Apure basin, which covers some 12,000 sq. km. of sparsely populated land, was known until recently for the quality of its agriculture and cattle. It was believed that it contained only relatively minor accumulations of oil (like Silvestre and Sinco). The discovery of large medium crude fields in the area (La Victoria, Guafita) has significantly altered this perception in recent years: Corpoven believes that these and other related structures hold about 1.4 billion barrels of recoverable light and medium crudes.[5]

In terms of possible crude deposits as yet undiscovered, it is fair to say that the probability of oil being struck in mainland Venezuela in the future still exists: the country has nearly 350,000 sq. km. of sedimentary basin onshore, out of which only a fraction (10–15 per cent) has been assigned for exploration (although, needless to say, the chances of finding a field even remotely approaching the Bolívar Coastal in size are infinitesimal). These unexplored areas are not necessarily 'virgin' territory and even mature areas are far from having been explored in their entirety. For instance, PdVSA estimates that, in the Maracaibo basin, an area equivalent to 25 per cent of the total surface of the basin remains to be explored (CEPET, 1989, v. 1, p. 96). The Falcón basin seems to be an area of very limited appeal; the likelihood of its containing a new giant field is slim indeed. In contrast, the location of some formations of the lightly explored Barinas-Apure basin, would seem to augur well for future exploration there. For instance, Guafita (the largest structure in the area) is apparently a geological extension of Colombia's giant Caño Limón field. Exploration-wise, however, PdVSA's highest hopes in the mainland remain pinned to the medium gravity crude fields of the Oriental basin. El Furrial-Musipán reserves are, as of 1990, in the neighbourhood of 1.6 billion barrels, but PdVSA expects to raise the proved and probable reserve estimate by 5.5 billion barrels by 1994, and it believes that the reserve total might eventually exceed 8 billion barrels. PdVSA, understandably, likes to put its money where its hopes are: in its 1991–6 budget, the states of Monagas and Anzoátegui were made the recipients of as much as 40 per cent of the exploration budget of $1.6 billion (*O&GJ*, 19 August, 1991, p. 16).[6] Furthermore, since the discovery of new sources of light and medium crude oil is an objective of the highest priority for both PdVSA and the government, this exploration budget may find itself exempt from the cuts inflicted on PdVSA's 1991–6 budget in 1992.[7]

According to the common wisdom of the oil industry, however, the most promising area for future exploration in Venezuela lies off-shore. To date, only 30 per cent of the 100,000 sq. km. extension of the Venezuelan continental shelf has been explored in any way (CEPET, 1989, v. 1, p. 153). In particular, the Gulf of Venezuela – an area 200 km. long and 125 km. wide, and which has scarcely been drilled – has considerable potential.[8] Unfortunately, the issue of boundary demarcation (Colombia and Venezuela have yet to agree on its division) has always prevented intensive exploration operations from being carried out there.[9] Of late, however, the Venezuelan government has floated the idea that it might consider embarking upon a major exploration effort in the area, Colombian protests notwithstanding.[10]

Historically speaking, however, off-shore exploration has so far been a source of keen disappointment for PdVSA, whose expectations of possible findings have been consistently wide of the mark by a considerable margin. At the onset of its involvement with off-shore drilling, PdVSA ex-president Alfonzo Ravard expressed a great deal of confidence about possible discoveries, saying the company expected to find 'reservoirs of not less than 10 billion barrels of predominantly light oil, and possibly

40 billion [sic.]' (Niering, 1980, p. 193). Unfortunately, all the oil reservoirs that PdVSA has found off-shore have been commercially non-viable (CEPET, 1989, v. 1, pp. 152–3).[11]

Not all of Venezuela's future sources of oil supplies will come from new findings; fuller development of its known oil accumulations through the application of enhanced recovery methods can be expected to make available substantial volumes of crude. (According to Servello (1983, p. 141), only about 34 per cent of Venezuela's original oil in place in the proven areas – without taking into consideration deposits in the Orinoco Oil Belt – has been subjected to secondary recovery processes. See also CEPET, 1989, v. 1, p. 195). From 1950 onwards, a great deal of effort has gone into trying to increase the amount of recoverable crude in Venezuela's oilfields. This effort has met with a reasonable degree of success. For instance, the average recovery factor associated with primary production methods in light/medium crude oilfields is around 29 per cent. With the use of water and gas injection, this number jumps to upwards of 40 per cent (CEPET, 1989, v.1, p. 194). In the heavy oilfields of Lake Maracaibo, only about 11 per cent of the oil in place can be recovered through conventional primary techniques. The search for a way to boost the recovery factor in this area has resulted in the development of a very successful two-stage steam soak–steam drive technique. The use of steam soak on its own can bring the recovery factor in any given field up to 25 per cent, approximately. But if steam drive is applied subsequently, the average recovery factor increases to around 32 per cent. The steam soak method, in particular, has proved to be remarkably efficient, since it yields about 40 barrels of oil for every barrel used for steam generation. The reason for this efficiency, according to Arnold Volkenborn (former head of Maraven), is that the reservoirs 'are basically of a very unconsolidated nature and reservoir compaction results in pressure maintenance and oil displacement. This somewhat unusual compaction feature is extremely important and made effective in combination with steam injection. The result is low-cost oil often cheaper than primary means (*O&GJ*, 4 February, 1980, p. 76). Steam injection, in contrast with steam soaking, is far less efficient: it yields only about 3 barrels of oil for every barrel of oil burned for steam generation.

The reader will note that, so far, we have made no mention of what is arguably the most important hydrocarbon province in Venezuela; namely, the Orinoco Oil Belt. However, such is its significance that a chapter of the book will be exclusively devoted to it.

3.2 Crude Reserves: The Politics of Accountancy

Venezuela's proved reserves of crude oil, according to official figures, amount to roughly 59 billion barrels. This is equivalent to almost 6 per cent of the world's total crude reserves; it makes Venezuela sixth in the list of countries with the greatest oil reserves, while giving her an impressive reserves/production ratio of 69:1, at 1991 estimated production levels (see Table 3.2).

Table 3.2: Crude Oil Reserves by Country. Billions of Barrels. 1990.

Country	*Reserves*	*Reserves/production ratio**
Saudi Arabia **	257.5	113
Iraq ***	100	132
Kuwait **,***	94	237
Iran	92.86	76
Abu Dhabi	92.2	130
Venezuela	59.1	70

* At 1990 production levels
** Excluding Neutral Zone
*** At pre-Gulf War production levels
Source: *O&GJ*, 30 December, 1991, pp. 48–49.

Venezuela's reserves break down by type into: extra heavy crude 27.702 billion barrels; heavy crude, 15.044 billion barrels; medium crude, 8.823 billion barrels; light crude, 5.995 billion barrels, and condensates, 1.475 billion barrels (*O&GJ*, 14 January, 1991, p. 37). One should stress, however, that this breakdown can be very misleading, because it is based on a classification system for crudes developed by the ministry of energy and mines of Venezuela. This system divides crudes, according to their API gravity, thus (CEPET, 1989, v. 1, p. 323):

Light crudes : From 30.0° to 40.0° API inclusive
Medium crudes: From 22.0° to 29.9° API inclusive
Heavy crudes: From 10.0° to 21.9° API inclusive
Extra-heavy crudes: Less than 9.9° API inclusive

These criteria, however, differ widely from those used by the world oil industry. One can illustrate this with a simple example. By Venezuelan standards, Mexican Maya crude (22.5° API), would be considered a 'medium' crude, but it is probable that numerous refiners would strenuously object to this classification. After all, for many in the refining business, Maya is the epitome of a heavy, 'difficult' crude, calling for very sophisticated upgrading facilities for its processing.[12]

Arguably, though, any possible misunderstanding regarding the quality of Venezuela's reserves should be dispelled after seeing that 72 per cent of the total consists of crudes of under 22° API gravity. However, the composition of this 72 per cent is, in itself, a further (and major) source of confusion. At the heart of this confusion lies the idea that the oil industry has formed regarding the contents of the Orinoco Oil Belt. This area, due to the physical characteristics of the hydrocarbons in place and the technological requirements necessary to exploit them fully, has come to be considered (quite rightly, it would seem) as a *non-conventional* crude oil source, in the same category as, say, the Athabasca tar sands or the Colorado shale deposits. Recoverable reserves from such deposits – precisely because of their non-conventional nature – are normally assumed to be excluded from countries' reserves statistics.[13] Thus, it would seem reasonable to assume that, in the same way as crude from the Athabasca tar sands does not find its way into the reserves statistics of Canada, Orinoco crude is excluded from Venezuelan reserve statistics.[14] This view, however, is completely mistaken; in fact, crude from the Orinoco oil constitutes approximately one-third of Venezuela's total proved reserves.

As Figure 3.1 shows, the level of Venezuelan reserves grew modestly until 1986, when reserve figures for year-end (55.5 billion barrels) revealed an increase of 26 billion barrels relative to the close of 1985. This dramatic jump reflected the inclusion in the statistics of about 18 billion barrels of so-called 'easily

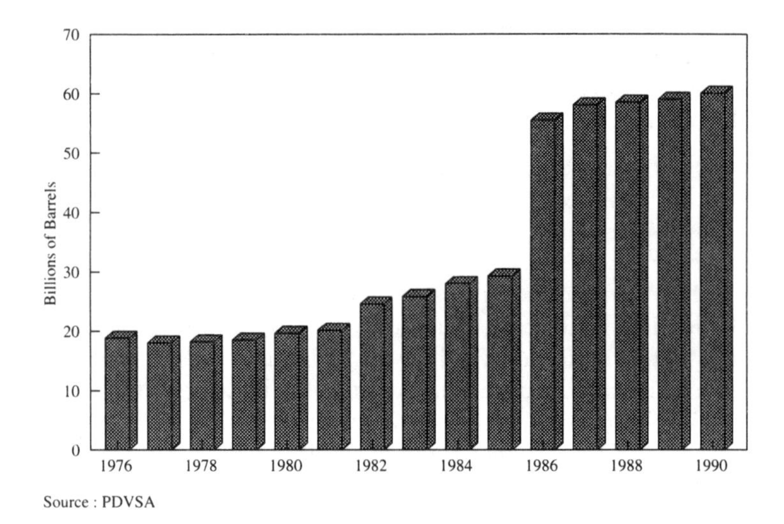

Figure 3.1: Evolution of Venezuelan Crude Oil Reserves.

recoverable' reserves of extra-heavy oil found in the Orinoco Oil Belt.[15] The breakdown of the 1986 reserve additions is as follows: 4 billion barrels of light and medium crudes; 4 billion barrels of heavy crudes from traditional production areas (Lake Maracaibo); and 18 billion barrels of Orinoco heavy crude.[16] The president of PdVSA at the time, Juan Chacín, justified this recalculation by arguing that the government's previous reserve figures were too conservative. Chacín emphasized that the new calculation was 'in keeping with international norms'[17] (*O&GJ*, 12 January, 1987), presumably in order to assuage any fears that Venezuela might be conjuring reserves from her sleeves.

It is a worthwhile exercise to examine these international definitions and concepts, in order to see whether Chacín's claims stand on firm ground. The first question is whether the hydrocarbons found in the Orinoco Oil Belt qualify as heavy crude (and are therefore worthy of inclusion in reserves statistics). According to international common usage, conventional heavy oil reserves are those which can be recovered to some degree through the natural drive forces of the reservoir, and which can be subjected to primary production by cold pumping. In contrast, non-conventional oil deposits (like oil sands) will only yield production by the application of some external force such as steam stimulation or surface mining (Millard, 1983, p. 163). This means that the conventional nature of a crude is, in the end, a function not so much of its API gravity, but of its viscosity, because it is this feature which determines whether the hydrocarbons from a given reservoir are recoverable at a commercial rate through a well in their natural state.[18] Hydrocarbons in the belt are extremely viscous at ambient temperature but, luckily for the Venezuelans, the Orinoco deposits are characterized by quite high reservoir temperatures. This favourable characteristic reduces the viscosity of the hydrocarbons-in-place, making possible the extraction of petroleum liquids without the use of steam drive or mining (this stands in marked contrast to the case of the Athabasca tar sands). Therefore, it is difficult to disagree with Martínez when he says that 'hydrocarbons in the belt are crude oils, [as opposed to natural tars] because their dynamic viscosities at original reservoir temperatures and atmospheric pressure, on a gas-free basis, range from 2,000 to 7,000 mPa.s' (1987, p. 131). Since 'even the heaviest crude in the area is producible by primary, conventional means . . . from the exploitation standpoint, this crude can be considered conventional' (*O&GJ*, 4 February, 1980, p. 72); hence, the inclusion of Orinoco crudes in the Venezuelan reserve tallies seems justified. One should point out, naturally, that the recovery factor associated with primary production methods in the Orinoco would be quite low – around 3 per cent, according to Tiratsoo (1984, p. 373) – but that is another problem.

From a purely economic standpoint, however, Venezuela's quantification of her reserves is a bit more suspect. As Brown points out, '[the] official existence [of reserves] depends on expected finishing cost. Reserves will not be officially counted unless it appears economically profitable to produce them. Large volumes of oil are known to exist which are not included in official reserves because it would cost more to produce them than what the oil would sell for' (1989, p. 178). In essence, this means that the development and extraction costs for a given crude deposit must be taken into account before it is actually classified as reserves, because as Adelman has written, 'reserves are not found, only oil-in-place is found; reserves are developed' (Adelman, 1972, p. 25). Or, in other words, oil-in-place that cannot be developed economically is better left out of reserves statistics; the mere knowledge that oil is to be found in some quantity in the subsoil is not a sufficient condition to justify its being included in a nation's reserves statistics.[19]

PEMEX, Mexico's national oil company, has been accused of succumbing to the temptation of artificially enlarging the magnitude of Mexican reserves by including in them approximately 11 billion barrels of crude from the Chicontepec zone, whose recoverability at this moment in time is suspect (*PIW*, 16 December, 1991, p. 5). Chicontepec's very difficult geological structure[20] makes massive capital expenditures in infrastructure an absolute necessity; hence, this basin would be a prohibitively expensive place to develop.[21] Many see the possibility of PEMEX recouping the investment necessary to bring Chicontepec on stream as a remote one indeed. Therefore, it has been argued, Mexico should have waited for Chicontepec's development to be economically feasible before including this dubiously recoverable oil in reserves statistics (Baker, 1992, p. 36).

The same kind of case can be made for the Orinoco Oil Belt. Even though Venezuela has a sizeable infrastructure in place in the Orinoco area (unlike Mexico in Chicontepec), it is still doubtful whether, at present, crude from existing wells can be exploited (that is, produced *and* sold) as crude. Even if one assumes the oil can be economically extracted, there is no reason to believe that anybody will buy if it has not been extensively upgraded beforehand. Indeed, history has shown that there is *every* reason to believe the exact opposite would be true (i.e. buyers would be noticeable by their absence).[22] And, as Brígido Natera, PdVSA's president in 1986, wisely put it, 'there is no reason to produce or develop . . . this type of oil if you cannot sell it' (Zlatnar, 1986, p. 28).

The only way of making Orinoco crudes acceptable to conventional refineries would be to upgrade them. At present, the jury is still out on whether the economics of such an enterprise justify the tremendous investments necessary to put it under way.[23] According to the US General Accounting Office (GAO), PdVSA officials think 'it would be profitable to produce and upgrade . . . extra heavy crude on the basis of 1992 world oil prices and estimated production and upgrading costs'. Major international oil companies, however, have taken a much less sanguine view of things. According to a GAO survey of 22 major oil companies, a number of them were 'not interested in producing heavy and extra heavy crude because they believed heavy oil activities were economically marginal, they lacked the experience of producing . . . extra heavy oil, and/or they considered the technology for refining extra heavy crude to be commercially unproven' (*O&GJ*, 3 February, 1992, pp. 26–7).[24] This appraisal, of course, might be wrong. Nevertheless, one should never forget that, as the GAO said, the 'economic viability [of such a project] would be very sensitive to changes in world oil prices and processing costs' (ibid.).

Two conclusions can be drawn from this analysis of the composition of Venezuelan crude reserves. First, the Venezuelan nomenclature system for crude reserves can lead many people to think that the country's so-called 'conventional' crude resource base is larger than it actually is. Secondly, a sizeable proportion of these reserves constitutes crudes which, given their extraction, processing and refining requirements, do not seem to be candidates for economic exploitation at the moment nor, therefore, for inclusion in the reserves statistics. In fact, most of the Orinoco reserves are probably better defined by the OPEC-coined category of 'semi-proven reserves', i.e. reserves 'which geological and engineering information indicate, with a fair degree of certainty, to be recoverable from a tested reservoir, or [reserves which are] technically recoverable, to a high degree of certainty, but the exploitation of which is deemed uneconomic' [presumably 'at present'] (Tiratsoo, 1974, p. 336). Granted, there is little reason to doubt that 'non-conventional' crude oil [like that in the Orinoco] will, with time, become 'conventional' (World Petroleum Congress, 1984, p. 7) but, until this happens, Venezuela's reserve accounting system must be regarded as rather questionable.

The reasons behind Venezuela's overstatement of reserves are not difficult to fathom. First, oil reserves are one of the many criteria (others being production capacity or past production) determining a given country's OPEC production quota.[25] Thus, other things being equal, large oil reserves can give a country a better negotiating position *vis-à-vis* other members of the cartel.[26] Secondly, the size of a country's oil reserves has a great bearing upon its standing in world financial markets. Larger reserves usually mean better credit-worthiness, and increased attraction to potential investors.[27] For a given country, all this means that if it is to 'err' at all in its reserve appraisal, it will be better off by doing so on the high side. Finally, the size of Venezuela's reserves is a variable of great importance when considering the country's relationship with its most important client from a historical point of view, the USA. Venezuela has always emphasized (publicly and noisily) its reliability as a supplier of crude oil to America. This exercise in public relations aims to soothe US sensibilities about excessive dependence on imports (sensibilities which might give rise to measures like imposing importation quotas or massively subsidizing US syncrude production), while at the same time magnifying Venezuela's advantages as a crude supplier compared to other countries, located in politically unstable areas of the globe. Thus, big reserves are essential (or, certainly help) to maintain the image of Venezuela as a producer that will keep the USA reliably supplied with oil for decades to come. However, for all of Venezuela's claims, it is very clear to people who understand geology (and can therefore interpret correctly the data on productivity, life expectancy and wealth of her fields) that Venezuela is *not* a Caribbean Saudi Arabia (*EE*, September 1988, p. 5).

3.3 Crude Oil Production: the Implications of Maturity

Commercial quantities of oil were first discovered in Venezuela in 1914. By 1929, a production of 137 mb of crude per year was enough to make it the second most important country in the world in terms of total oil output (just behind the USA), a position which it would keep until 1961, when it was superseded by the Soviet Union. After 1929, its oil production continued to increase at a steady rate – from 1926 to 1947, Venezuelan production exceeded that of the entire Middle East combined (Lieuwen, 1985, p. 192) – peaking in 1970, at a level of just over 3.7 mb/d.[28] Output began to decline thereafter, reaching its lowest level in 1987 (see Table 3.3). However, over the

Table 3.3: Venezuelan Crude Oil Production.

Year	*Million Barrels per Day*
1970	3.708
1973	3.364
1978	2.165
1979	2.356
1980	2.167
1983	1.791
1984	1.724
1985	1.669
1986	1.664
1987	1.592
1988	1.658
1989	1.731
1990	2.118
1991	2.341

Sources: *IPE, O&GJ*, various issues.

period 1988–91, and as a result of the 1990 Gulf crisis, it has risen almost to the total production capacity.

A look at the Venezuelan production profile (Table 3.4) reveals a slide in the average percentage growth rate of crude production which is typical of declining, mature producers, such as the USA. Even the production from crude reservoirs found after 1986 has been unable to arrest this trend. The gradual and irreversible decline of the Maracaibo oil province is the main culprit behind Venezuela's overall production slide, as a look at some numbers from the largest oil-bearing structures in the area (Lagunillas, Bachaquero, Tía Juana, Lama,

Table 3.4: Average Percentage Growth Rate of Venezuelan and US Oil Production.

Period	*Venezuela*	*USA*
1960–65	3.66	1.67
1965–70	1.39	3.67
1970–75	−7.87	−1.66
1975–80	−5.88	−0.33
1980–85	−5.86	0.79
1985–90	2.21	−3.26

Sources: API, 1992; MEM, 1989.

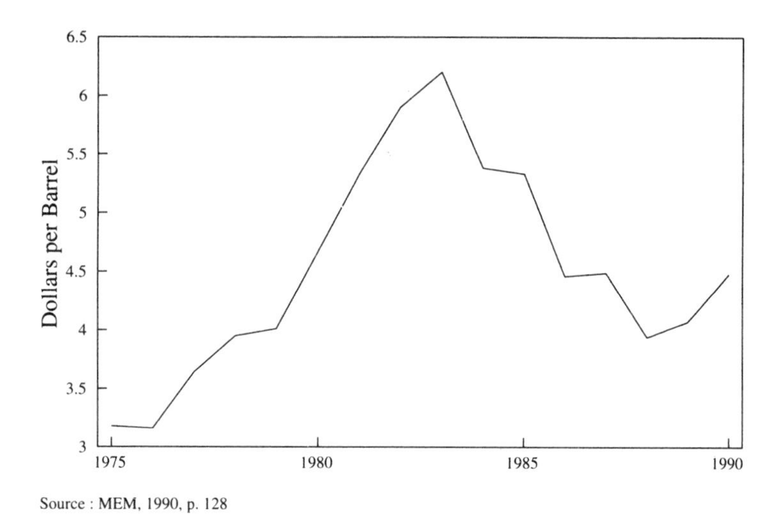

Figure 3.2: Venezuela. Oil Production Costs, 1976–90.

Lamar, and Centro), will reveal. In 1970, when Venezuela's overall production was in its heyday, these fields accounted for 2.7 mb/d or 72 per cent of the total. By 1979, total production capacity had fallen to no more than 2.4–2.5 mb/d, and production from these fields hovered around 1.5 mb/d, a 44 per cent drop from 1970 levels. During 1990, output from these giant fields came to just 1.13 mb/d (or 53 per cent of the Venezuelan total), notwithstanding the array of enhanced recovery projects which PdVSA operates in the area.

This declining trend has had a very important consequence: for PdVSA, maintaining static production levels has come to require increasing effort; thus, for instance, the number of producing oil wells in Venezuela has grown (with all the financial outlays that this implies), but lower production yields per well have not resulted in increases in field output proportional to the increases in the number of wells.[29] Also, reservoir depletion (advancing at a rate of 17 per cent annually) and well depressurization have forced PdVSA to make growing and extensive use of costly secondary and tertiary oil recovery methods (see CEPET, 1989, v. 1, pp. 194–8).[30] However, as Figure 3.2 shows,[31] PdVSA's overall production costs have come down quite noticeably since the mid-1980s. This declining trend is due to a variety of factors, and reveals the complexity of the cost picture of the Venezuelan upstream sector. First, since the decline roughly coincides with the coming on-stream of the medium crude fields found in the second half of the eighties, one can say that the changing character of the country's production slate had a positive bearing on costs, given that heavy crude is more difficult to extract, and hence more expensive to produce than light crude. However, the downturn in the cost curve also coincides with the end of the 1979 oil boom. This means that the large amount of funds at the disposal of PdVSA during this time of affluence could also have contributed to pushing its production costs up.[32]

Venezuela's production capacity also suffered a nosedive after the halcyon days of the late 1960s. During 1970–77, it fell to 2.454 mb/d, from a previous level of 4 mb/d. These developments were a legacy of the sagging exploration which characterized the Venezuelan upstream during the late 1950s (a consequence of the government's policy of granting no new concessions). The falloff in output, capacity and reserve additions accelerated drastically in the early seventies, once the major oil companies understood that it would not be very long before the nationalization of Venezuelan oil took place. Furthermore, at this time the technical expertise of the state oil company, CVP, was still very much at an embryonic level, so that no help came from these quarters in order to offset the shortfall. Indeed, as Coronel (1983, p. 135) put it, 'the greatest accomplishment of the oil industry during the period 1976–9 was to keep production capacity at a level of about 2.2 mb/d.' After exploration by the nationalized industry finally kicked off in earnest in 1979, PdVSA managed to push capacity to approximately 2.5 mb/d when the goal was 2.8 mb/d (Niering, 1982, p. 86). Paradoxically, by the time this second target was finally approximated, PdVSA found itself under great pressure to shed its excess capacity. The reason for this was that a production capacity nearly 1 mb/d greater than what the country could actually produce (given its OPEC quota) proved to be a very contentious issue in a Venezuela shaken by the tremors of the international debt crisis, and in the grips of an austerity programme. Many politicians argued repeatedly that sheer common sense called for saving money by forsaking the luxury of idle capacity. PdVSA officials objected to these suggestions vociferously. Idle capacity, they argued, permitted Venezuela to respond to market fluctuations and strengthened its position within OPEC (*International Petroleum Encyclopedia*, 1986 ed., p. 121). In the end, their view prevailed. Events in the Middle East (the 1990 Iraqi invasion of Kuwait) were to vindicate the foresight of the bureaucrats who had advocated a retention of idle capacity. Granted, when the call on this excess capacity was placed by the market in August 1990, it took PdVSA three months to increase the total oil flow to 2.5 mb/d (from a level of 2 mb/d), because the de-mothballing turned out to be more difficult than expected. None the less, one should not forget that during the Gulf crisis, Venezuela was one of the countries whose upstream rose to the occasion, and produced significant incremental volumes of crude.

The 1990–91 Gulf crisis has made increasing both production and production capacity once again a matter of overriding concern for PdVSA. In the eyes of the Venezuelans, the crisis dramatically underscored Venezuela's advantages as an oil provider for oil-consuming nations desiring stable, long-term supplies of oil; to wit: its location (it lies outside the politically volatile Middle East); its political system (it is a reasonably stable and democratic country); its oil policy (its flow of oil has never been stopped by regional wars or crises, or embargoes); and, finally, its oil endowment (it possesses the largest proved reserves of crude outside the Persian Gulf). In top Venezuelan oil circles, it is a widely held opinion that the country's inability to reap benefits commensurate to the importance of these advantages has been due to the fact that it has always been the odd one out in an Arab-dominated OPEC. Thus, 'Venezuela has long felt that it is outnumbered [within OPEC] by the Gulf States, and that its quota [has] failed to reflect its reserves' (*The Economist*, 9 March, 1991, p. 72). But the political fallout associated with the Gulf crisis has radically altered the balance of power within OPEC. Saddam Hussein's ill-advised foray into Kuwait opened a deep rift between the different nations of OPEC's Arab bloc, a rift which has come to separate the Arab countries which Saudi Arabia perceives as 'friends' from those it considers 'foes'. This gap is much harder to bridge than the previous one dividing 'price doves' from 'price hawks'. Its existence means that Saudi Arabia is no longer willing to contemporize with the demands of some of its less powerful Arab brethren, in the name of Arab solidarity (as post-January 1991 OPEC meetings have shown quite clearly). Bearing in mind a scenario where production capacity is *the* variable that determines a country's power and leverage within OPEC (and hence, its production quota[33]), to the detriment of previously significant aspects like population, revenue needs or revenue absorption capacity, PdVSA has planned to hike its production potential to more than 3.6 mb/d, by year-end 1995 and 5 mb/d by the year 2000 (in other words, the company is aiming for future production capacity parity with Iraq and Iran). In this way, the Venezuelans seek to strengthen their position within OPEC and *vis-à-vis* their biggest customers. The success of these quite ambitious plans, however, is far from assured. Committing large amounts of money to a cause, as Venezuela's failure to boost production capacity in the late 1970s shows, is no guarantee of success. In any case, the precarious finances of the government have made it impossible for PdVSA to even try to achieve this goal: the cuts in its 1991–6 production expansion budget mean that a major stroke of luck will have to occur if PdVSA's crude production is to reach the 5 mb/d mark by the century's end.

3.4 Main Crude Types: The Not-so Light, the Very Heavy and the Extraordinarily Heavy

Venezuela's first commercial oil strike (Mene Grande) produced a heavy, 17.8 API crude. Ever since, heavy crude has always been centre stage in the country's long oil production history (in fact, Venezuela has accounted for over 50 per cent of the cumulative heavy oil production in the world to date). Venezuela has certainly been a case where the Freudian aphorism, 'childhood is destiny', has come true with a vengeance. There are numerous Venezuelan crude streams, derived from various combinations of crudes from different fields. These streams are shown in Table 3.5. where it can be seen that among this wide range of crudes, the heavy, sour varieties predominate.[34]

Table 3.5: Venezuelan Commercial Crude Streams.

Name	*°API*	*Fields*	*Type**
Lagocinco	35	Lagocinco	P
Lagomar	32	Lagomar, Mara, Sibucara	P
Lagotreco	31	Lagotreco	P
Ceuta	30	Ceuta	P
Cabimas	20.6	Cabimas	N
Menemota	19.6	Mene Grande, Motatán	M
Lagunillas	15	Lagunillas	N
Bachaquero BCF-13	13	Bachaquero, Tasajera	N
Tía Juana Heavy	11.6	Tía Juana	N
Laguna	11.2	Laguna 11	N
Pacón	31	La Paz, La Concepción	P
Mesa-Oficina	28	Oficina, Mesa	P
Anaco-Wax	41	Aguasay, Tucupido	P
Leona	23	Various fields	M
Merey	16	Various fields	N
Lagomedio	33	Centro	P
Lamar	33	Lamar	P
D.Z.O.	29.5	Lago, West Coast	P
Colón	39	Casigua	P
Corredor	29.8	Corredor	P
Boscán	10.6	Boscán, Los Claros	N
Silvestre	23.1	Various fields	N
Pilón	12.5	Temblador, Pilón, Jobo-I and diluent	N
Caripito	18.5	Tucupita, Pedernales	N
Jobo/Morichal	12	Jobo-II, Morichal, PICV-Jobo, Cerro Negro and diluent	N
Tía Juana Light	31.5	Tía Juana	P
Bachaquero BCF-17	17	Urdaneta, Bachaquero	N
BCF-21	21.9	Tía Juana, Urdaneta	P
BCF-24	24	La Rosa	P
Cretáceo	42	Cretáceo	P
TJ-102	25.8	Tía Juana 102	N
Tía Juana Medium	25.7	Tía Juana	P
Furrial	28	El Furrial	P

*P: Paraffinic; N:Naphthenic; M: Mixed composition
Source: CEPET, 1989, v. I, pp. 324–5

The Venezuelan production slate has always been disproportionately skewed towards lighter crudes (considering the size of its heavy crude reserves). This is completely logical, given the adequate availability of lighter grades in Venezuela up to the

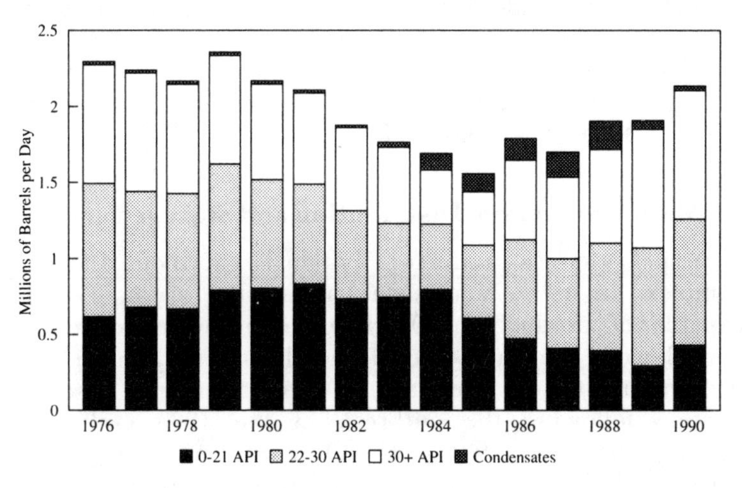

Source : PDVSA

Figure 3.3: Venezuelan Crude Oil Output by Gravity.

early 1970s, and the undesirable characteristics of heavy crudes (not only are they costlier to produce, but also the sulphur and metals in the crude damage boiler tubings, and sulphur in stack gas is a serious atmospheric pollutant; heavy crudes also have a higher tendency than their lighter counterparts to adhere to tubing walls and this produces tremendous back pressures during pumping operations). However, after the early 1970s Venezuelan light and medium crude reserves began to shrink, because of the very intensive exploitation to which they had been subjected. This resulted in a fall in light and medium crude output, which had as its inevitable corollary a progressively heavier Venezuelan crude slate. In 1977, for instance, heavy crudes (i.e. crudes under 22° API) accounted for 30 per cent of output, but by 1984 they were responsible for 44 per cent of the total (see Figure 3.3). Following the discoveries of new light crude fields in 1986 (which many characterized as purely fortuitous), the proportion of total Venezuelan outcome accounted for by heavy crudes fell to 25 per cent of total output. However, with the incorporation of Orinoco crudes in the Venezuelan reserves, the distribution of production according to the reserves of each type of crude became more skewed in favour of light crudes than ever before. In 1990, for instance, although crudes with a gravity greater than 22° API made up just 28 per cent of the country's reserves, they accounted for over 81 per cent of Venezuela's 2.1 mb/d average output (*PIW*, 4 November, 1991, p. 8).

3.5 Main Production, Distribution and Storage Facilities

Venezuela's crude production system is enormous and very complex. In 1989, Venezuela had a total of 26,505 wells capable of production. Of those, 9,557 were active and 16,948 were shut-in. Some of the shut-in wells have since been reactivated, as a result of PdVSA's decision to raise production following the outbreak of the 1990 Gulf crisis (*O&GJ*, 19 August, 1991, p. 17). Other existing production facilities include 877 production stations, 164 gas compression plants, 11 gas processing plants and 1,414 water injection plants.

Venezuela possesses 9,000 km. of major oil pipelines, as well as nearly 30,000 km. of flux pipelines, 700 km. of diluent transportation pipelines (vital for extra-heavy oil production) and 11,000 km. of pipelines transporting gas for use in gas lift operations (CEPET, 1989, v. 1, p. 321). The oil transportation capacity of this pipeline network in 1989 was 8.8 mb/d, with 1.2 mb/d concentrated in Eastern Venezuela and 3.8 mb/d in the Maracaibo region.[35]

Venezuela has 21 crude tank farms distributed throughout the country. Total storage capacity in these facilities as of year-end 1989 totalled 15.9 mb. These tank farms are used to rid the crude of water and salt before sending it to a refinery or terminal; they also function as centres for effluent treatment. Storage capacity in Venezuela's oil terminals comes to 40.5 mb for crude and 78.4 mb for products. In addition, PdVSA controls 47 mb of storage capacity outside of Venezuela.[36]

There are 16 oil terminals in Venezuela, most of them with state-of-the-art shipping-handling installations. These include two riverine terminals on the Orinoco, export terminals in the big refining complexes of Amuay, Cardón, Puerto La Cruz and El Palito, and eight export terminals in Lake Maracaibo (including Puerto Miranda, the newest in the country). It should be noted that all the terminals located in Lake Maracaibo suffer from an important handicap; namely, that the navigation channel which gives access to the main body of the lake is quite shallow. This means that it has to be dredged constantly, in order to assure the safe passage of big tankers; this operation, its bothersome aspects apart, is quite expensive (Coronel, 1983, p. 101).

In 1991, PdVSA had a sea-going fleet of 19 vessels, with a combined tonnage of 768,000 dead weight tonnes (dwt), and an average age of 13 years. The tankers (eight for dirty products, seven for clean products, two for asphalt and two for LPG) used to be operated by the three integrated subsidiaries of PdVSA,[37] but they are now handled by PdVSA Marina. PdVSA Marina has planned a major replacement/expansion programme for the tanker fleet, to be carried out throughout the 1990s. This programme contemplates an expenditure of $1.3 billion through 1995 on new tankers. The first step of this programme has already been accomplished with the purchase of eight 85,000 dwt double hulled crude tankers from Hyundai Heavy Industries, at a cost of $61.9 million each. Hyundai will deliver the first tanker in 1993 (it will also be the first Venezuelan double hull ship), with additional deliveries every six months thereafter. The deal was financed by Mitsubishi, and it was spread out over 12 years with a two-year grace period. In addition to the eight crude carriers, PdVSA Marina will acquire two 20,000 dwt asphalt carriers, a lubricants carrier, five 35,000 clean tankers, four 60,000 dwt tankers for Orimulsion and two LPG tankers. These acquisitions are planned to give the company the capability to transport up to 30–40 per cent of its exports, compared to the current level of 12–13 per cent (*O&GJ*, 26 August, 1991, p. 36).

3.6 Reactivation of Marginal Oil Fields: Wringing Blood out of a Stone?

Although the 1990–91 Gulf crisis gave the expansion of crude production capacity a top priority billing in the agenda of the world's oil producers, it has also, by virtue of an ironic twist of fate, made this goal extremely difficult to achieve. This is because the capital needs associated with the reconstruction of the oil sectors of Iraq and Kuwait are bound to clash with the financial requirements of producer countries who want to expand their production capacity (that is to say, all of them). In a world which is currently undergoing a period of marked capital scarcity,[38] and where a rise in oil prices (which would provide an incentive for more investment in the sector[39]) does not seem to be in the offing, investment in oil seems to have become a zero-sum game. Hence, in the future, it is reasonable to say that for every nation that achieves the goal of capacity expansion, there will be others which will be left out in the cold. The opening up to foreign investment of the oil sectors in the republics of the Commonwealth of Independent States only serves to complicate matters further, by introducing into the scramble for oil money, qualified manpower and technology an actor (Russia) whose needs can only be described as colossal.

Venezuela's policy-makers realized, at a relatively early stage in the development of the Gulf crisis, that the speed at which the country could expand its oil production capacity was a matter of crucial importance. Therefore, so as to get a head start on its competitors, PdVSA came up with a scheme which, in theory, was to allow a rapid expansion of Venezuela's crude production by 150,000–200,000 b/d, at a minimum cost to the company. Basically, the plan envisaged bringing on-stream once again marginal fields which, due to Venezuela's need to comply with her OPEC quota, or to their adverse operating economics, had been idled for the better part of 30 years. Reactivating the fields would have been uneconomic for PdVSA's subsidiaries, given their high overhead costs, but the company thought that smaller firms might be able to operate them profitably. The company made it clear that no equity interest in the fields would be granted, and that only service deals were contemplated. This was a convenient arrangement because it dispensed with the need of Congressional approval for the project.[40] And, although by its very nature, the project was better suited to specialized engineering and drilling firms, a number of integrated majors expressed interest in it, seeing it as the prelude to future equity deals in the country (*PIW*, 8 April, 1991, p. 3).

All the fields considered in the plan (see Appendix 2) were located in the Oriental and Falcón basins, where reserves ranged from condensate to heavy crudes. The cumulative production of the fields was put by PdVSA at 1.438 billion barrels, with remaining proved reserves of 357 mb and remaining probable and possible reserves of 1.169 billion barrels (*O&GJ*, 19 August, 1991, p. 15).

The details of the model contracts have been kept under wraps ever since the programme was announced, but it seems that the fees paid to the operators will relate directly to the volume of crude produced. The contracts will be long term, and operators are expected to provide upfront capital two years before production actually begins. Operators will be allowed to drill appraisal and development wells in their assigned fields, but drilling beyond pay zones producing when the fields were last operating will require the negotiation of a new exploration and production contract (*O&GJ*, 24 February, 1992, p. 43).

In perspective, PdVSA's expectations of a moderate capacity rise at a reasonable cost appear to have been a bit optimistic, because to date the reopening of the marginal fields has not been an unqualified success. Only 19 companies and groups out of the 88 that initially acquired data for the licensing round bothered to enter bids. Major oil companies, in particular, were conspicuous by their absence in the definitive tendering rounds.[41] More serious than this, and even more deleterious to the company's plans, was the fact that only 5 out of 9 offered areas drew a bid.[42] This made a downwards revision of PdVSA's production estimates imperative.

The first major milestone in the reactivation programme came on 23 June, 1992, when PdVSA and the ministry of energy and mines selected the winning bids for the five areas which drew offers. The winning companies were Shell de Venezuela, for the Pedernales unit; the Benton Oil & Gas Co./ Vinccler partnership, for the Urocoa, Bompal and Tucupita units; Lingoteras de Venezuela, for the onshore Falcón and West Guárico units; and Teikoku, for the East Guárico unit (*PON*, 23 June, 1992, p.3).

Once the fields had been awarded to the winners, the programme immediately ran into a barrage of heavy political flak. The Venezuelan congress, which had been expected to be a quiet party to the deal, raised its voice in protest, and branded it as the beginning of the denationalization of oil. Shortly after this, Lagoven's talks with Shell on the matter of investment took a sudden turn for the worse, when Shell insisted that its contract include a clause which stated that any dispute between the contracting parties be settled in an international court. Lagoven's refusal to accept this proviso (which would have required congressional approval, with all the political squabbles that this would imply) resulted in the cancellation of Shell's operating concession.[43] Fortunately for PdVSA, BP appeared on the scene, and pronounced itself ready to fill the void left by Shell.[44] BP also accepted local arbitration and other Venezuelan laws. Apparently, BP's acquiescence has a lot to do with the fact that it expects to receive the right to drill below the original producing horizons of the field, in the hope of striking new reserves of light and medium crude (*PIW*, 23 November, 1992, p.5).[45]

It should be remembered that the possibility of companies being able to drill beyond the specified contractual depth was not contemplated in the original plans for marginal field reactivation. However, the results of the first round of bidding were so discouraging that PdVSA included this new clause in its offering of a further 20 inactive fields, with proven reserves of 1.2 billion barrels (*PIW*, 23 November, 1992, p.5). Whether this action will result in a more lively bidding round for the fields on offer, remains to be seen.

3.7 Conclusion: Perspectives for Oil in Traditional Production Areas

The main production areas in Venezuela, because of their age, are in need of large investments. A major part of PdVSA's short-term strategy, in fact, deals with the replacement of oil reserves in these traditional production areas; depletion makes it necessary for PdVSA to replace around 80,000 and 100,000 b/d of crude every year, just to be able to sustain a 2.1 mb/d production level constant.[46] This implies major expenditures, even if the company's reserve addition costs are a quite reasonable 57 cents per barrel.[47]

To offset somewhat the imbalance in the composition of its crude reserves, PdVSA plans to add 10 billion barrels of light (i.e with a gravity greater than 22°API) crude to its proven reserves in the period 1991–5. To accomplish this, the company will undertake 50,000 line km. of seismic surveys and drill more than 130 wildcats. On the face of it, Venezuela has a resource base ample enough to reward these efforts: in addition to its 59 billion barrels of proven reserves, PdVSA puts the country's probable reserves at 85.5 billion barrels and possible reserves at 55.6 billion barrels, for a combined total of 200.1 billion barrels. Furthermore, according to the *Oil&Gas Journal*, '[this] does not include 271 billion barrels of extra heavy crude and bitumen the company believes it can recover from the Orinoco Oil Belt' (*O&GJ*, 14 January, 1991, p. 36). These numbers seemingly contradict a 1983 DOE/EIA study which said that there was a 95 per cent probability that 12 billion barrels of conventional oil sources remained to be discovered in Venezuela, and that the probability of finding 38 billion barrels of reserves was only 5 per cent. Since this study was published, Venezuela's oil reserves have experienced a dramatic increase, but its conclusions have not been really challenged, since the great majority of the reserve increases were due to reappraisals of Orinoco reserves, *which the DOE study did not consider* (DOE/EIA, 1983, p. 3). Thus, given the previously explained confusion regarding Venezuela's definition of reserves, it would be advisable to approach *O&GJ's* figures with some scepticism, because there is a good chance that Orinoco crudes make up a part of its estimate (and a sizeable part, at that).

Before the budgetary cuts announced in the second quarter of 1992, PdVSA planned to increase its capacity to a level of 3.6 mb/d by 1996, and its actual production to a level of 2.2 mb/d (3.2 mb/d had been floated as another possible target, but this was perceived as a bit far-fetched). The key element in this plan was to get the light and medium crude output up from 1.7 mb/d to 2.2 mb/d – its probable plateau – and to increase heavy from 500,000 b/d to 1 mb/d (*EE*, December 1991, p. 18).[48] These plans have now been abandoned, and the company is expected only to hold its output potential constant, at a level of about 2.8 mb/d. The reopening of marginal fields, unfortunately, is not likely to help PdVSA significantly (at least not to the degree that the company would have liked), because it seems very clear that major international oil companies will not be lining up at the door in the near future, if service contracts are all the company has to offer them.[49]

Notes

1. The supergiant and giant oilfields in the Maracaibo basin are: Bolívar Coastal, Boscán, Centro Ceuta, La Concepción, La Paz, Lamar, Las Cruces, Los Claros, Mara, Mene Grande, Urdaneta; in the Barinas basin: Guafita, Silvestre, Sinco and La Victoria; and in the Maturín sub-basin: Aguasay, Boca, Chimire, Dación, El Furrial, Guara, Jobo, Jusepín, Las Mercedes, Leona, Limón, Mata, Merey, Morichal, Musipán, Nardo, Nipa, Oficina, Oritupano, Oscurote, Oveja, Pedernales, Pilón, Quiriquire, San Joaquín, Santa Ana, Santa Rosa, Soto, Temblador, Trico, Yopales and Zapatos.
2. With the important exception of the Boscán formations, which date from the Eocene and Oligocene, and have very heavy oil in place (CEPET, 1989, v. 1, p. 205).
3. Known in Venezuela as 'menes'. Some of these even created quite large asphalt lakes (like Guanoco).
4. The most important oilfields in the group are Aguasay, Boca, Chimire, Dación, Ganso, Guara, Güere, Güico, Guario, Leona, Limón, Lobo, Mangos, Mata, Miga, Nardo, Nigua, Nipo, Orión, Oritupano, Oscurote, Oveja, Silla, Soto, Trico, Yopales, Zapatos, Zorro, Zumos (Tiratsoo, 1984, p. 370).
5. As of 1990, the area's proved reserves were about 600 mb.
6. PdVSA's investments in these states will not be devoted entirely to light and medium crude exploration, however. A part of it will go to exploration in the Orinoco Oil Belt, which also lies within the borders of these states.
7. Especially after the very promising results of a major seismic survey of the area became known in July 1992. Lagoven said that, based on an initial interpretation of the results of the survey, it estimates that an area east of El Furrial may contain up to 4 billion barrels of reserves (*PIW*, 6 July, 1992, p. 12).
8. Richard Nehring, of the Rand Corporation, wrote in 1980 that, among the unexplored or lightly explored potentially oil-bearing provinces of the world, there were probably no more than a dozen with a decent probability of exceeding 6 billion barrels in crude reserves. Among these he included the Gulf of Venezuela. The others were the Beaufort Sea/ Mackenzie Delta (Canada), the Malvinas Basin (Argentina–U.K.), the Lena-Anabar and East Siberian Sea provinces (Russia) and the East China Sea province (Nehring, 1980, p. 175)

9. The Colombian side of the Gulf of Venezuela (including the eastern portion of the Guajira peninsula) was blanketed with offshore concessions, never explored because of a storm of official Venezuelan protests, which began in 1967. The Venezuelans claimed that the concessions granted by Colombia extended beyond her territorial water jurisdiction. The whole issue revolved around the location of Los Monjes, a cluster of uninhabited isles which Colombia had officially ceded to Venezuela. Based on the fact that Los Monjes is quite close to Colombia, Venezuela has claimed jurisdiction over a large portion of the Venezuelan Gulf.
10. There could be rich pickings for PdVSA in another offshore area as well: the undeveloped Posa-112 field, discovered in the Gulf of Pariá in 1958, lies on the same structural trend as Soldado, an important offshore field in Trinidad and Tobago (Tiratsoo, 1984, p. 371). Also, in 1989, Lagoven proved up oil in the onshore Boquerón structure, believed to be part of a structural complex embracing other parts of the Gulf of Pariá, and which could contain, according to PdVSA, at least 20 billion barrels of oil (*Euroil*, April 1992, p. 27). See Appendix 1.
11. Although commercially exploitable accumulations of natural gas have been discovered in the CarCpano offshore basin.
12. One consequence of Venezuela's rather 'loose' definitions of different hydrocarbon typeswas the infamous 'condensate dispute' with Kuwait and Saudi Arabia in 1988. This conflict arose when, in a tight international oil market situation, these states realized that, after the introduction of crude production quotas in OPEC (which explicitly excluded gas liquids production), 'Venezuelan output of condensate [had] started to rise sharply using a definition that [was] much looser than general industry standards' (*O&GJ*, 4 July, 1988, p. 13). According to Saudi Arabia and Kuwait, Venezuela's official definition of condensate made it impossible for OPEC's auditors to distinguish between light crude and condensate, and this would have a disastrous effect on the cartel's production discipline. The dispute came to a head after they threatened to reclassify some of their light crude streams as condensate, and withdraw them from the quota system (which would have taken as much as 1.3 mb/d of crude out of the quota system, assuming that no other producers had followed suit). Arturo Hernández Grisanti, oil minister of Venezuela at the time, rejected the possibility of changing his country's definition of condensate, arguing that it had been in force since 1971. However, when the Arab members of OPEC stepped up the pressure, Venezuela finally agreed to have its 'condensate' production included as part of its OPEC oil production quota (see *O&GJ* 4 July, 1988, pp. 12–14). Viewing the whole question impartially, it seems that Saudi Arabia and Kuwait had a good point. Their definition of condensate was similar to the generally accepted oil industry standard (liquid hydrocarbons with more than 50° API gravity at 15.56°C); by Venezuelan standards (which define condensates as hydrocarbons occurring in gaseous phase in underground reservoirs at original reservoir conditions, liquid at atmospheric pressure and temperature, and having a gravity higher than 40.2° API at 15.56°C) some crudes, called volatile oils,

could easily be considered condensates (these types of oil are found, for instance, in the Albuskjell and West Ekofisk fields of the Ekofisk production system).

Volatile oils are found in reservoirs where the temperature is rather close to, but still slightly below, the critical temperature (or cricondotherm) at which the hydrocarbons in the reserve assume a gaseous phase. The cricondotherm defines a level above which hydrocarbons in two phases cannot coexist. However, as one gets nearer to the cricondotherm, identifying the physical state of the hydrocarbons (gaseous or liquid) *in situ* becomes progressively more difficult (the term 'dense phase' is often used to describe hydrocarbons under these conditions). The problem with the Venezuelan definition of condensates is that, under certain physical circumstances, it further blurs the tenuous distinction between light oils found in dense phase and condensates.

The essential difference between volatile oils and other crudes lies in their respective compositions. In the case of volatile oils, since reservoir temperatures are near the critical point, the composition of the gas phase and liquid oil phase is rather similar. The gas phase in volatile oil accumulations is unusually rich in intermediate and heavy components, while the liquid phase is very rich in light components.

13. For instance, British Petroleum's *BP Statistical Review of World Energy*, June 1991, states that its estimates of world oil reserves by country do not include oil from shales or tar sands.
14. Witness in this sense the opinion of the leading oil-market journal *Petroleum Argus* (27 January, 1992, p. 19): 'Total [Venezuelan] conventional reserves (excluding the Orinoco Belt) are pegged at about 60 bn bls, of which about 40 bn bls are conventional heavy crude.'
15. This does not imply that Orinoco crude was not included in Venezuelan reserves before 1985, because it certainly was, as these lines from the *Oil & Gas Journal* (22 August, 1983, p. 72) show: 'Venezuela's reserves at year end 1982 were 24.6 billion bbl . . . That includes Orinoco reserves of 4.3 billion bbl.'
16. At year-end 1987 there was another hike of 2.5 billion barrels, a result of major discoveries of light and medium crude in eastern Venezuela and in Lake Maracaibo (*O&GJ*, 18 January, 1988). The period 1988–90 has seen additions amounting to a further 2 billion barrels.
17. According to PdVSA, definitions and concepts utilized in Venezuela for the estimation of proved reserves of oil and gas are the same which are conventionally used by the world petroleum industry. These definitions and concepts are based on the statements of the Association of World Petroleum Congresses, the American Association of Petroleum Geologists (AAPG) and the American Petroleum Institute (API). (CEPET, 1989, v.1, p. 199)
18. This is in keeping with the recommendations of the World Petroleum Congress, which state that viscosity should be the main criterion to define tar sand oil and heavy oil, with API gravity to be used as a proxy only when viscosity measures are not available (*Platt's Oilgram News*, 18 February, 1982, p. 3). Thus, crude oil is the portion of petroleum that

has a viscosity of less than or equal to 10,000 millipascal-seconds (mPa.s) at original reservoir temperature and atmospheric pressure, on a gas-free basis (World Petroleum Congress, 1984, p.5). According to the WPC, the distinction between crude oil and natural tar is based on viscosity rather than on whether the fluid in a reservoir is recoverable by conventional means because, in this way, the classification depends entirely on the characteristics of the fluid, and not those of the mother-rock (ibid. p. 30). This argument is not entirely convincing, because it conveniently overlooks the fact that viscosity is, among other things, a function of reservoir temperature (which means that the classification can never entirely depend on the characteristics of the fluid).

19. This is the linchpin upon which hangs the question of whether an area of an oil reservoir can be considered as 'proved reserves'. In the words of the API, proved reserves are those which can be 'reasonably judged as *economically* productive on the basis of available geological and engineering data' (Tiratsoo, 1974, p. 334. Italics mine).
20. The oil wells drilled in Chicontepec have found relatively tight and discontinuous turbidite pay zones with high water saturation (averaging 44 per cent), low solution gas to oil ratios, relatively heavy crude, clay cementation, low porosity (averaging about 7 per cent), and low permeability (averaging less than one millidarcy). (DOE/EIA, 1983(a)).
21. According to the *International Petroleum Encyclopedia* (1991 ed., p. 119), full development for Chicontepec would involve around 10,000 wells; production would be around 100 b/d per well. However, the USA Department of Energy calculations put even this figure in doubt, saying that 'although the initial recovery production per well may range up to 150 barrels of oil per day, the average sustained daily production is . . . approximately 40 barrels per day' (DOE/EIA, 1983(a), p. 45)
22. In the early part of the 1980s, the fact that Venezuela would be running out of medium and light crudes within a few years began to dawn upon Venezuelan oil policy-makers (although this event, of course, was forestalled by the 1985–6 discoveries). Well aware of the marketing difficulties posed by heavy oil, PdVSA embarked on a drive to line up long-term purchasers for very heavy crudes on the basis that the importer would construct upgrading facilities in exchange for guaranteed supplies. The first of these deals was struck with the French company Elf Aquitaine. Under the terms of the deal, Elf was supposed to build a pilot very deep conversion refinery near Marseilles at a cost of 1 billion 1986 dollars (the whole sum to be covered by Elf), in order to process 2 mt a year of 10° API Boscán crude. This pilot refinery was supposed to have been the precursor of more ambitious developments in Venezuelan crude upgrading, but in the end Elf found the economics of Boscán refining so daunting that this collaborative venture was stillborn. The reader would do well to bear in mind that the quality of Boscán, especially in terms of critical aspects like viscosity or API gravity, is actually superior to that of the average extra-heavy Orinoco crude!
23. PdVSA officials estimate marginal production costs for Orinoco crudes to be around $2.80 per barrel. The cost of upgrading this crude could be as

high as $13 per barrel (*EC*, 20 March, 1992, p. 8). But these costs should not be taken as indicators of the breakeven level for the enterprise, since they do not consider capital investment, which for a 100,000 b/d upgrading plant (no pipelines, wells, etc.) would be $2.5–3 billion at the least (*O&GJ*, 3 February, 1992, p.26).

24. A number of foreign oil companies (Veba, Elf, Eni, Conoco, Mobil, BP, Chevron, Amoco, Mitsubishi/Mitsui, Total, Itochu – formerly C.Itoh – and Shell) have signed letters of intent to help Venezuela develop its tremendous heavy and extra-heavy crude production potential. However, one should not try to see too much behind all this because, after all, signing letters of intent comes easy. What will happen when PdVSA tries to make one of these companies allocate some money for the ventures, is anybody's guess. In the opinion of *WPA* (27 January, 1992, p. 17), 'the slow pace at which companies that have already signed are committing to specific projects is already fuelling concern in some political circles and raising questions about the level of financial commitment implied by the strategic alliances.' Indeed, given the state of the oil industry in the latter half of the nineties, this author would be surprised if any of these companies commit a substantial amount of money in these ventures.
25. See the report 'The OPEC Quota Issue Revisited' in pages 37–44 of the January–February 1992 issue of *Global Oil Report*.
26. Venezuela took this conclusion to its limits in May 1992, when oil minister Alirio Parra formally asked OPEC to take its reserves of heavy oil – which it put at 700 billion barrels, a much higher figure than the 200 plus billion usually mentioned – into account for the calculation of its OPEC quota (*EC*, 29 May, 1992). One would expect that other OPEC members would greet this suggestion with some scepticism, but Parra was confident that the proposal would not affect Venezuela's status in the organization in the medium to long term.
27. A recent Ex-Im Bank–PEMEX agreement contemplates the bank guaranteeing loans tobe raised by Pemex through bond issues in international capital markets. The purpose of this plan is to raise money to finance a drilling programme to be undertaken in Mexico by American companies. And, as *PIW* (16 December, 1991, p. 5) has noted, 'oil reserves play a key role in convincing bond buyers to take up the offering.'
28. Ironically enough, Venezuela lost its position of most important oil exporter in the world (which it had held since 1946) to Iran and Saudi Arabia in 1970, the year of its greatest production.
29. In 1975, Venezuela had less than 21,000 oil wells, which could produce 2,654 mb/d. In 1986, producing the same volume of crude required more than 26,000 wells. *International Petroleum Encyclopedia*, 1986, p. 121.
30. The fact that many analysts believe that a sizeable proportion of the $16.2 billion allocated to production expenditures in the much criticized PdVSA expansion plan for 1991–6 will be used merely to sustain production at 1990 year-end levels, is a logical corollary to all this. While estimates as to the exact proportions vary, there seems to be no debate as far as the soundness of the basic premises of the argument is concerned

(*O&GJ*, 14 January, 1991, p. 37). What a journalist wrote in 1977 still holds true, in the sense that it is necessary for PdVSA to run as fast as it can in order to stand still; to make any progress it has to run even faster (Kim Fuad, quoted by Philip, 1982, p. 470).

31. This figure does not plot marginal production costs (which would be significantly lower than those depicted), because allowances for overheads have been factored in.
32. In Venezuela, the technically rated (t.r.) capacity and the actual production capacity of a given well are not necessarily identical; thus, in order to get a well to produce its t.r. capacity, PdVSA must spend *x* amount of money on that well. In times when funds are widely available, the company might put money into every well in its system (i.e.'gold plating'), to get them all up to their t.r. capacity. It will also demand more of certain goods (the services of engineering firms, well specialists and the like), which will push the price of these goods up. Naturally, these factors will lead to higher production costs. In times of economic contraction, the process goes into reverse: the company becomes more selective on the wells it would like to spend money on, and its demand for the aforementioned goods comes down.
33. The words of former PdVSA president Andrés Sosa Pietri, show that Venezuela believes this to be true: 'If PDV wants to be a big player, it must increase [crude] production' (*PIW*, 29 October, 1990, p.2).
34. See also Appendix 3.
35. The pipelines in the Maracaibo region move approximately 70 per cent of the Venezuelan production (CEPET, 1989, v.1, p. 411).
36. Distributed like this: Borco, Grand Bahamas (20 mb); Bopec, Bonaire (9 mb); CDT, Curaçao (18 mb). The first two facilities are wholly owned by PdVSA, while the third is operated under a lease agreement (*WPA*, 27 January, 1992, p. 18).
37. Corpoven owned two clean tankers, with a combined tonnage of 59,000 dwt. Lagoven had the greatest number of ships (three clean tankers, five dirty tankers and two asphalt carriers; combined tonnage: 470,000 dwt). Maraven operated the two LPG tankers, as well as two clean tankers and three dirty tankers (combined tonnage: 283,000 dwt.). (CEPET, 1989, v. 1, p. 445)
38. As *Petroleum Argus* has put it: 'The world has to face up to the prospect of capital shortage. The deficit on the USA budget is now expected to increase. Probably the USA payments deficit will increase with it and the inexorable need of the USA to import capital. The people who supplied the capital to offset these two deficits in the past are not able to supply it now. The Japanese banks have a precarious balance sheet. They have a shortfall of assets owing to the fall in real estate. The Germans have their hands full financing the east, and the Arabs have to pay for the Gulf War and for the repair of war damage. The US is going to have a problem. It may be relied upon to solve this problem and somehow or other to find the capital it needs. *The solution has to be at the expense of the rest of the world which will go short*' (22 July, 1991, p. 1. Italics mine).

39. To quote from *Argus* again: 'Both [oil companies and producing country governments] pin their hopes on a rise in the oil price to get the capital and the exploration effort moving again. But the price rise may not materialise. It all depends on the Saudis. The Saudis are in the unique position of being able to finance their own development. They may not give their support to a rise in price that would enable their competitors to finance development too' (7 October, 1991, p.1).
40. Although in June 1992, some Venezuelan congressmen complained that Congress had not been properly informed about the marginal field reactivation plan by PdVSA, and that the contracts would require prior congressional review to determine whether they fall under the provisos of article 5 of the nationalization law (projects coming under this article require approval by a joint session of Congress). *PON*, 30 June, 1992, p.2.
41. The *Oil&Gas Journal* (16 March, 1992, p.36) reported on the Venezuelan reaction to this not very encouraging report: 'Government officials had hoped a better response to the program because 88 companies or combines initially had shown interest in bidding for the contracts. However, PdVSA expressed satisfaction that the bidding had attracted some major oil companies.' By 'some major oil companies', one is supposed to understand Shell and BP, which bid for the same field, and perhaps Repsol, which together with four other companies submitted a bid for 20 fields.
42. The following production units received no bids: Quiriquire, Jusepin, West Falcón.
43. Maraven's deal with Lingoteras also fell through, because the winning company was unable to find a financial backer for its investment. This means that only two of the field awards resulted in contracts being signed ahead of the deadline.
44. BP was a runner up when Shell won the bidding for the service contracts.
45. Since the mid-1980s, deep drilling to the west of the Pedernales unit has turned up over 9 billion barrels of proven and probable reserves of light and medium crudes.
46. Now that PdVSA's production has grown, it would be logical to expect that the reserves addition needs will be greater.
47. According to PdVSA's production co-ordinator, Francisco Prada, the company's finding costs are just over 13 cents per barrel. They are expected to rise to about 20 cents per barrel in a few years' time, and to reach 80 cents per barrel by the end of the century (*PE*, June 1992, p. 39).
48. Venezuela's heterodox usage of certain terms can cause confusion as to the exact effort that was needed to achieve these designs. Since PdVSA's definition of productive capacity takes into account all possible flows (including those from shut-in wells awaiting workovers), PdVSA's available capacity in 1992 was actually 2.5 mb/d, and not 2.8 mb/d (the figure officially stated by the company). Thus, even if the 1992 year-end target of 3 mb/d had been met (and this would have entailed arresting a decline of over 600,000 b/d, while adding 160,000 b/d), this capacity would still have been 'potential' productive capacity (*PIW*, 23 March, 1992, p.5).

In other words, to get actual capacity to 3 mb/d, PdVSA would have required an additional 300,000 b/d increment, on top of all the others.)

49. To put it clearly: 'International oil companies don't like to work for fees . . . We want to invest risk capital and get oil in return' (*PE*, March 1992, p. 8).

4 THE ORINOCO OIL BELT

4.1 An Ocean of Oil

Nature can be said to have been somewhat biased when she determined the distribution of hydrocarbons around the world. Oil deposits, for instance, are only found in a very small proportion of the world's sedimentary basins, with great quantities of oil concentrated in an even smaller number of giant fields.[1] The same can be said to be true of extra-heavy crude and tar sands deposits: supergiant accumulations account for the vast majority of the world's reserves of these hydrocarbons. The seven largest extra-heavy crude and tar deposits in the world contain 98 per cent of the world's heavy oil in place (equivalent, by the way, to the oil in place in the world's 264 giant crude oilfields). The heavy oil deposits of western Canada (892 billion barrels), for instance, are larger than the proved oil reserves of the entire Middle East (Demaison, 1978, p. 203).[2] However, even Canada's tar sands are dwarfed by the vast size of Venezuela's Orinoco Oil Belt, arguably the biggest hydrocarbon province in the world.

The Orinoco Oil Belt covers an area of 54,000 sq. km. and contains more than a trillion barrels of oil in place (although, oddly enough, there are no significant surface indications of petroleum anywhere in the belt). The belt is 700 km. long and 80 km. wide, and runs roughly in parallel with the northern bank of the Orinoco river. Hydrocarbons in the belt occur predominantly in tertiary non-consolidated sediments of deltaic origin, generally at depths of less than 920 metres. Average temperatures in the hydrocarbon reservoirs in the belt fluctuate between 50°C at 800 metres and 60°C at 1,000 metres. These reservoirs are mainly found in extensive, non-discrete, 'major producing areas', called Cerro Negro, Pao, Hamaca, San Diego, Zuata and Machete,[3] whose average oil saturation varies significantly. Thus, while Cerro Negro has saturation ratios of 85 per cent, San Diego and Machete, for instance, have values in the order of 80 and 71 per cent, respectively (Martínez 1987, p. 132). The main characteristics of the

Table 4.1: Orinoco Oil Belt. Reservoir Characteristics.

	Minimum	*Maximum*
Net oil sand thickness (feet)	20	500
Depth (feet)	900	4,500
Pressure (pounds per square inch)	400	2,000
Temperature (° Fahrenheit)	100	150
Oil saturation (per cent)	70	80
Porosity (per cent)	27	32
Gas-oil ratio (SCF/STB)	10	150
Sediment type	Nonconsolidated sands	

Source: Servello, 1983, p. 153.

Table 4.2: Orinoco Oil Belt. Data on Three Development Areas.

	North Hamaca	*Cerro Negro*	*San Diego*
Barrels of oil in place (billions)	30	200	177
Average net oil sand thickness (feet)	100–200	200–400	100–250
Average depth of pay (feet)	3,500	2,500	3,000
Average °API gravity	9–12	8–9	9–12
Cold production (b/d)	N.A.*	180	140–700
Production after steam injection	N.A.*	390	1,000
Extent of area (square kilometres)	1,500	2,800	2,200

* Not Available

Sources: Servello, 1983, p. 157; Martínez, 1987, p. 134.

reservoirs and crude oils in the belt can be seen in Tables 4.1, 4.2 and 4.3.

The discovery well of the Orinoco Oil Belt (La Canoa-1), drilled in 1936, revealed enormous layers of sedimentary sands, saturated with heavy (7°API) crude. However, the high sulphur content, metallic residuals and extremely high viscosity of the crude made it difficult to produce and commercialize, and as a result, the company abandoned the well. During the three decades following this initial discovery, only 58 wells were drilled in the belt, notwithstanding the fact that, as Servello (1983, p. 153) points out, 'almost every

Table 4.3: Orinoco Crude Characteristics

Gross heat of combustion (Kcal/Kg)	$9.5x10^3$–$9.9x10^3$
Gravity (°API)	4–17
Carbon (percentage of weight)	84.0–86.0
Hydrogen (percentage of weight)	10.0–10.8
Sulphur (percentage of weight)	2.0–5.0
Nitrogen (percentage of weight)	0.6–0.8
Oxygen (percentage of weight)	0.75–0.85
Ash (percentage of weight)	0.10–0.13
Vanadium (PPM)	200–1,000
Nickel (PPM)	50–150
Iron (PPM)	10–15
Sodium (PPM after treatment)	40–70
Conradson carbon (percentage of weight)	15.0–17.0
Flash point, °C	120 min.
Pour point, °C	21 min.

Sources: EE, March 1988, p. 16; Servello, 1983, p. 155

exploratory well drilled within [the area] showed impressive heavy oil sands'. Thus, even though it had revealed signs of harbouring staggering amounts of hydrocarbons, the belt remained very much 'an oddity, or a geological quirk' (Martínez, 1987, p. 125) up until the late 1960s. The decade of the 1970s brought about a radical change in the status of the Orinoco Oil Belt, at least within Venezuelan circles. This change was noticeable, first of all, in the name bestowed to the belt. As Tugwell wryly remarks, '[Venezuelan] government spokesmen [began] to refer to it by the more optimistic name of *faja petrolífera* [instead of *faja bituminosa* (or tar belt), the name prevalent until then]' (1975, p. 139). Secondly, after all the years of neglect it had experienced (as far as investment was concerned), the belt became the focus of an extraordinary appraisal and exploration effort by the new-born PdVSA.[4] In 1978, the company divided the belt into four sectors (from east to west: Cerro Negro, Hamaca, Zuata and Machete). Each of these sectors was assigned to one of PdVSA's four operating subsidiaries (Lagoven, Meneven, Maraven and Corpoven, respectively),[5] which in turn were ordered to complete a thorough geological and sedimentological analysis of the region (investigating the structural pattern definition in the area, types of geological traps, possible quantities of oil in place, location of the most important accumulations and potential production areas), as well as an evaluation of the best recovery and upgrading techniques for the Orinoco crudes. The magnitude of PdVSA's five-year effort – which implied, among other things, drilling 669 wells totalling 643,000 metres, doing 595 reservoir tests, and covering 54,000 km. by airborne magnetometer survey – was calculated by Martínez (1987, p. 128) to amount to 2,500 man/years, at a cost of 2.6 billion *bolívares* ($615 million).[6] Needless to say, these numbers stand in stark contrast to the dearth of investment which had been the lot of the Orinoco Oil Belt before 1978. But this statement begs the question: why was it that the marginal 'tar belt' of the late 1960s became the vitally important oil belt of the early 1970s?

The roots of this transformation can be traced back to the decline of reserves and of oil production in traditional producing zones in Venezuela, which started to become apparent during the early 1970s. The government realized that, in the long term, this decline – compounded with growing domestic demand for oil products – would shrink the exportable volume of oil quite dramatically. Given Venezuela's unsuccessful attempts to diversify her economic base, and the government's overwhelming dependence on oil exports for its revenue, these trends were quite rightly perceived as ominous. Furthermore, with the realization that the widespread use of enhanced recovery techniques would, at best, only brake the production decline, many top policy-makers in the country concluded that, in the future, Venezuela's crude oil export margin would necessarily have to come from undeveloped areas (the most attractive among which was the hitherto dismissed ocean of tar lying under the alluvial plains of eastern Venezuela).

The period of sharp increases in the price of petroleum during the early seventies gave the development of the Orinoco province a further boost. The effect of these increases on the Venezuelan economy was enormous and far-reaching, as Chapter 8 of this study shows. The revenue windfall caused by these price increases led President Carlos Andrés Pérez to embark on a public expenditure programme of unprecedented magnitude and ambition, which saw the construction of steel mills, hydroelectric power stations, shipyards, and aluminum smelters, as well as the introduction of massive welfare programmes. All of this had the avowed objective of realizing the long cherished dream of 'sowing Venezuela's oil' (*sembrar el petróleo*, a phrase coined by conservative politician Arturo Uslar Pietri in 1936). Rather unfortunately for Venezuela, the increase in world oil prices coincided with the aforementioned decline in oil production and reserves. Although this fall did not matter in immediate revenue terms – precisely because of the price increases – 'the extent to which it reflected the country's moribund reserve conditions boded ill for future production and exports' (McGowan, 1990, p. 919). This was a matter of concern because oil exports, as usual, were expected to be the means by which the massive debt acquired by the government in order to finance Pérez's mega-projects was to be liquidated. It also became clear to the government that it would not be doing much sowing *in the future* unless it managed to increase quite substantially the country's productive capacity (and thus enlarge the exportable surplus). Consequently, plans were drawn up to take Venezuelan crude production capacity to a level of 2.8 mb/d.[7] However, achieving such an increase without tapping the Orinoco Oil Belt resources was rightly perceived as an impossibility.[8]

PdVSA's initial strategy for the exploitation of the Orinoco Oil Belt was to follow a two-pronged approach to crude production in the area, by implementing two projects of very different nature and complexity. The first project (entrusted to Lagoven), whose aim was to produce upgraded crude from Cerro Negro, in the south of the state of Monagas,[9] was to make use of the most advanced oil production and processing technology available at the time. The centrepiece of this project was to have been a production and upgrading complex which would have permitted the production of 140,000 b/d of upgraded 25° to 30° API crude, with a 1 per cent sulphur content, and which would have been virtually free of vacuum bottoms. The upgrading plant would have functioned along the following lines: 70,000 b/d of light oil diluent were to be constantly cycled from the plant to the producing wells, where the diluent would be blended with heavy oil (this scheme was expected to facilitate oil handling and dehydration operations, as well). This crude mix would then have been run

Table 4.4: DASM Project Estimated Investment. Billions of Dollars.

Item	*Amount*
Upgrading Plant	3.8
1,600 production wells, equipment	1.4
Other production facilities	0.5
Infrastructure (roads, etc.)	0.5
Pipeline system	0.3
Lifting terminal	0.2
Total	6.7

Source: Servello, 1983, p. 157.

through on-site atmospheric distillation units. Afterwards, delayed carbon rejection coking units[10] would coke the vacuum residual of this mix. Then, the coker distillates with the virgin distillates would have been hydrotreated, mixed to form synthetic crude, and sent by pipeline to Puerto La Cruz; this synthetic crude would have entered the world market at premium price over similar gravity crudes (which would have lacked the no vacuum bottoms bonus). All of the producing wells in the area (1,000) were to have been stimulated by steam soaking techniques; electrical power for the refining and producing functions was to be supplied by hydroelectric means and gas and coke would have been burned as fuel to generate steam (Servello, 1983, p. 158 and *O&GJ*, 15 September, 1980, p. 183). As Table 4.4 shows, the investment required to put this project in operation would have been massive. When the price of oil fell in the early 1980s, however, PdVSA was forced to reappraise its investment priorities, and thus the very costly DASM project was indefinitely postponed.[11] Whatever infrastructure had been put in place in Cerro Negro before the large-scale project was terminated has since been used to investigate the behaviour of the formation under high rates of production, the optimal spacing between wells, quality and degree of continuity of the sands, design and use of underground and surface equipment, and the control of well subsidence (CEPET, 1989, v. 1, p. 258), and more recently, for the production of the Orimulsion boiler fuel.

The second Orinoco development project (entrusted to Maraven) was far more conservative in scope than DASM.

Called Guanipa 100+, the project was intended to permit the production of 100,000 b/d of virgin crude by means of a blend oil scheme, using existing facilities in the south of the state of Anzoátegui.[12] The scheme would have functioned by mixing 'virgin' extra-heavy Orinoco crude downhole with lighter crude oils from fields in the north of the state. The resulting 16° API mixture would have been transported by existing pipelines to a terminal on the Caribbean Sea, ready to be exported. Although Guanipa 100+ was a rather straightforward project, nevertheless, for all its simplicity, it would still have had a rather daunting price tag attached to it, as Table 4.5 shows. In the aftermath of the 1986 price collapse, a cash-strapped Venezuela discovered that blending its lighter (and more valuable) crude oils with Orinoco crude contradicted economic wisdom quite flagrantly, and therefore, Guanipa 100+ followed DASM into oblivion.

Table 4.5: Guanipa 100+ Project Estimated Investment. Billions of Dollars.

Item	*Amount*
1,300 production wells, equipment	0.7
Other production facilities	0.4
Infrastructure	0.2
Total	1.3

Source: Servello, 1983, p. 158.

4.2 Orimulsion: the Key to the Orinoco's Riches?

PdVSA's decision not to proceed with the full-scale development of either Guanipa 100+ or DASM, did not spell the end of its subsidiaries' involvement in the Orinoco area. Rather, the Orinoco became a testing ground of sorts for Corpoven, Meneven and Lagoven. The three companies, with an eye to a more distant future, used the infrastructure they had in place in the area to evaluate different experimental recovery and transportation techniques for extra-heavy crudes. PdVSA president Brígido Natera justified the companies' behaviour thus: 'It will be a few years before we get into the Orinoco oil belt in a big way. [But] we will continue to learn more about how to handle it, better ways of treating it, and I am sure we will find a way to use it before the end of the century'

(Zlatnar, 1986, p. 34). In fact, Natera's words turned out to be prophetic, because as early as 1986, quite some time before the century's end, PdVSA happened to discover Orimulsion, a revolutionary new fuel which has the potential to transform the global power generation sector.

The discovery of Orimulsion resulted mainly from INTEVEP's research into the problems posed by the transportation of extra-heavy crude from the Orinoco. Usually, the flow of heavy crude oil through a pipeline is not possible if the crude is in its virgin state, because its great viscosity stresses pumps too much and thus limits the distance across which it can be moved. Therefore, upgrading the heavy oil in the field to pipeline quality is seen as a necessary step before any attempt to move the crude is made. A number of alternatives to achieve viscosity reduction exist: they range from the heating of pipelines, to diluting the heavy crude with lighter oils, or diluting it with natural gas condensates. However, in all but a few parts of the belt, none of these solutions was feasible. The area lacked the quantities of light oil and natural gas required for dilution processes, and pipeline heating, given the great distances which separate the area from the nearest port, and the extreme viscosity of the Orinoco crudes, would have been a prohibitively expensive proposition. Because of this, in the early 1980s, INTEVEP began to investigate two methods of heavy oil transportation which obviated the need for dilution or heating. The first, called core annular flow, was quite 'esoteric', in that it did not involve any viscosity reduction. It consisted of surrounding the oil in the pipeline by a water ring, reducing the pressure to values close to that of water, which facilitated pumping operations considerably. Although trials with this method confirmed that it was quite effective (flow rates of up to 34,000 b/d were recorded, and no problems developed during periods of standstill), they also confirmed that once the crude had been delivered to a marine dispatch terminal, the logistical problems would begin anew, because special tankers and reception facilities would be required to handle the crude. The second method aimed at viscosity reduction through the use of emulsions, and it was this line of research which eventually led to the development of Orimulsion when, in the course of investigations about the suitability of Orinoco crudes for fuelling operations in the oil industry, INTEVEP realized that the emulsions could be burned quite efficiently (McGowan, 1990, p. 922).

After this initial discovery, it did not take long for PdVSA to realize that the emulsions of Orinoco crude and water could be potential world beaters in the power station fuel market niche. In order to ascertain their suitability for power generation, it therefore commissioned a series of test firings in plants in Venezuela, using 150,000 tonnes of the product, while simultaneously 3,800 tonnes were sent out to sea to evaluate its transportability by tanker (*EE*, March 1988, p. 14). When these preliminary trials had been completed to the company's satisfaction, large-scale pilot commercial tests involving foreign power generators were begun. Among the companies PdVSA enlisted to test Orimulsion (the commercially patented name of the emulsions) were the Nagasaki Research and Development Centre (Mitsubishi), the Kreisinger Development Laboratory of Combustion Engineering, and Northern Engineering Industries International Combustion. Orimulsion, on the whole, did quite well in all the tests.[13] NEI's report, for instance, concluded that 'in all respects, the Orimulsion compared favourably with heavy fuel oil and [it] should prove an excellent fuel for use in generation boilers' (ibid., p. 15). One month later, Mitsubishi Heavy Industries made public results very similar to those obtained by the Scottish utility. At the end of 1988, and in view of the satisfactory results obtained from the pilot firings, PdVSA decided to launch its new product on a worldwide commercial scale.

Orimulsion is a product consisting of 70 per cent Orinoco extra-heavy crude (bitumen), 30 per cent water, and an emulsifying agent which stabilizes the mixture, and prevents it from separating into its original components. The emulsion retains its qualities so long as the product is kept away from other fuels and within a temperature range of 30–80°C. In terms of energy content, a pound of Orimulsion will give off 13,000 Btu worth of heat (steam coal will produce 12,000 Btu/lb and heavy fuel oil 17,400 Btu/lb). Another important characteristic of Orimulsion, and one which distinguishes it from pulverized coal, is that the particulate it holds in suspension is quite small. This means that Orimulsion combustion is complete at a relatively low temperature, thus conferring on the product a favourable NOx emission profile relative to that of coal, a distinct advantage in these times of heightened environmental awareness (EE, December 1991, p. 19). According to PdVSA, the fuel possesses many of the best characteristics of both heavy fuel oil and coal, without many of their disadvantages.[14] Orimulsion is also easier to produce than HFO (since it obviates the need for a refinery), and far, far easier to store and transport than coal. The main physical and chemical characteristics of Orimulsion are summarized in Table 4.6.

Table 4.6: Orimulsion Characteristics.

Gross heat of combustion (Kcal/Kg)	$7.0x10^3$–7.5x103
Carbon (percentage of weight)	59.0–60.5
Hydrogen (percentage of weight)	7.2–7.8
Sulphur (percentage of weight)	2.2–2.8
Nitrogen (percentage of weight)	0.43–0.58
Oxygen (percentage of weight)	0.53–0.60
Ash (percentage of weight)	0.07–0.10
Vanadium (PPM)	280–350
Nickel (PPM)	68–80
Iron (PPM)	12–17
Sodium (PPM, after treatment)	40–70
Conradson carbon (percentage of weight)	10–12
Flash point, °C	130 min.
Pour point, °C	2 min.
Orinoco crude content (percentage of weight)	72 ± 2
Water content (percentage of weight)	28 ± 2
Mean droplet size (um)	17 ± 3
Dynamic viscosity, mPas at 20° C	800–1200
Dynamic viscosity, mPas at 50° C	400–600

Source: *EE*, March 1988, p. 16.

At the moment, Orimulsion production takes place only in Morichal, located in the Cerro Negro area of the Orinoco. The production process is quite simple. First, water and a surfactant are injected downhole into the wells (which are stimulated by steam soak), where they mix with the crude and form a primary emulsion.[15] The resulting mixture of virgin crude, salt water and gas is then pumped to a flow station where it is dehydrated, degassed and desalted. The emulsion is then transported to the Morichal processing centre, where more water and surfactants are added, giving the product its final composition. From there, Orimulsion is transported to the Punta Cuchillo riverine terminal on the Orinoco, which, at the moment, handles all Orimulsion export activities.[16] According to the *Petroleum Times*, PdVSA's research 'has ensured that a product dispersal "contingency plan" is available in the event of spillage, either on land or sea' (4 May, 1990, p. 6).

Undoubtedly, one of the main attractions of Orimulsion is the ease with which power plants that burn liquid fuels can be converted to the new product, without losing the ability to burn HFO. According to Lagoven, the conversion process takes a mere six months. Also, in order to further enhance Orimulsion's appeal to the very conservative utilities, PdVSA has hinted that, in some cases, it would be willing to consider negotiating deals which would partially offset the conversion costs accruing to a power generator wishing to use the fuel (*PT*, 4 May, 1990, p. 6).[17] According to some, the characteristics enumerated by Bitor in its 'ideal customer profile' (i.e. customers should have water tube boilers, direct fired or fluidized bed boilers) would effectively rule out 'many, except very large, users' from using the fuel (*PT*, 4 May, 1990, p. 6). However, Bitor insists that Orimulsion's flexibility is such that it will be able to fully satisfy every potential customer, by means of tailor-made contracts.[18] For all its optimism regarding utility adoption of Orimulsion, however, Bitor has had to recognize that the opportunities for the fuel in the industrial sector are, at best, limited.[19]

Despite its many advantages, Orimulsion does have an important shortcoming: it does not really qualify as a green product, assertions to the contrary notwithstanding.[20] Granted, Orimulsion produces significantly less particulates than coal and its combustion leaves a minimum amount of unburnt hydrocarbon and no trace of acid smuts, a negative feature associated sometimes with oil-fuelled power generation (Reader, 1992, p. 32). Furthermore, plants with electrostatic precipitation facilities (which are recommended for Orimulsion burning) can reduce the emission of particulates to a very low level. Also, as noted before, its nitrogen oxide emissions are less than those associated with coal and similar to those of heavy fuel oil (and power plants can also use Orimulsion in low NOx burners, to ensure compliance with NOx limits set by the EC for 1996 onwards). Finally, according to BP, Orimulsion has a 10–15 per cent lower coefficient of CO_2 than coal (McGowan, 1990, p. 923; *EE*, 16 August, 1991). But its environmental credentials go no further than this. As can be seen in Table 4.6, the metals (vanadium, nickel, magnesium) content of Orimulsion is extremely high. This makes the product quite prone to induce corrosion, and to produce highly metallic ash (although the use of additives could probably suffice to counteract these two tendencies).[21] Furthermore, Orimulsion's sulphur dioxide emissions are even higher than those produced by 3.5 per cent fuel oil (although less than those produced by 2.5 per cent sulphur coal); in the main, it is this characteristic which puts it at a distinct disadvantage in the environmental sweepstakes. These disadvantages, of course, are not insurmountable, because 'wherever local environmental legislation limits sulphur dioxide emissions, various technologies exist, and more are being developed, to remedy this problem at various stages of the combustion process – before, during and after combustion' (Reader, 1992, p. 32). The fuel, for instance, can be subject to gasification. This permits the removal of the sulphur compounds before combustion of the gas.[22] On the other hand, sorbants (such as limestone) can be used to absorb the SOx during combustion; the resulting sulphur compounds can then be removed in an electrostatic precipitator. Finally, flue gas desulphurization after combustion can be achieved by installing sorbants between the boiler and the smokestack. However, the installation of this type of equipment – which is unavoidable for a utility using Orimulsion and wishing to comply with the provisos of, say, the US Clean Air Act Amendments of 1991 – can make Orimulsion's competitive edge *vis-à-vis* other fuels (notably natural gas) disappear, because, at the moment, utilities are keener on utilizing fuels that do not require contamination abatement equipment installed *ex professo* to burn them. In other words, the cost of modifying power plants to handle Orimulsion's high sulphur emissions, may end up restricting product sales to a few large firms which can afford to have scrubbers installed in their smokestacks.

On the marketing side, PdVSA's new subsidiary, Bitor has established a number of subsidiaries around the world, in order to further the commercial cause of Orimulsion. These include BP-Bitor,[23] which markets Orimulsion in western Europe, Bitor USA, Bitor International (for eastern Europe and North Africa), and Bitor Japan, a joint venture with Mitsubishi. Teaming up with strong foreign partners to sell Orimulsion has been seen as a convenient way of circumventing PdVSA's weaknesses in marketing (*EC*, 16 August, 1991).

BP-Bitor has been the most active of these subsidiaries, securing the biggest Orimulsion supply contracts so far.[24] It also has developed a very innovative pricing mechanism which, in keeping with PdVSA's contention that the product competes against coal, not fuel oil, sets the Orimulsion price on the basis of a differential against an indicator that BP-Bitor believe is competitive with coal. This indicator is a coal index, similar in conception to *McCloskey's Coal Index*, which is calculated using a basket of international coal prices (*WPA*, 10 June, 1991, p. 2). As of year-end 1991, Bitor had signed supply commitments worth 7.1 mt of Orimulsion per year (equal to 138.25 boe/d) with utilities from Japan, the UK, Spain, Canada, and Portugal (see Table 4.7).

Table 4.7: Examples of Orimulsion Contracts*

Country	*Million Tonnes p.a.*	*Begins*	*Duration*
UK (PowerGen Ince B)	1	1991	5 years
UK (PowerGen Richborough)	0.5	1992	5 years
Spain (Unión Eléctrica Fenosa)	1	1991	10 years
Japan (Kashima Kita)	0.6	1991	5 years
Canada (New Brunswick Power)	0.8	1992	20 years
USA (Texaco)	1.6	1992	25 years

* Excluding trials

Source: *WPA*, 15 April, 1991, p. 6; 22 June, 1992, p. 8.

Currently, Bitor is in a process of negotiation with other important power generators in Germany, Italy, the Netherlands, the UK and Denmark, a process which could lead to further commitments in the order of 10–12 mt of product annually.[25] These actions are perfectly in accordance with PdVSA's stated goal of achieving sales of 20 mt of Orimulsion by 1997 (and 40 mt by 2000). However, it should be noted that current contracts for Orimulsion are nearing Venezuela's production capacity. Therefore, major investments on the part of PdVSA are required before any new deals can be struck. In order to circumvent (at least partially) these onerous investments, PdVSA has proposed to its Orimulsion marketing partners (BP and Mitsubishi) the constitution of integrated joint ventures, which would allow these companies to produce their own Orimulsion for export (*PIW*, 20 April, 1992, p. 6). Although both companies expressed misgivings about Venezuela's political stability after the February 1992 coup attempt by disgruntled soldiers, Mitsubishi has said that PdVSA's terms are quite attractive. This is a very important development, as a result of the drastic budget revision of 1992 (which affected PdVSA's investment plans for areas other than its core businesses of producing, refining and marketing crude oil), the role which foreign companies could play in the development of Orimulsion has been vastly enhanced. Indeed, without the help of foreign investment, the promise of Orimulsion as a revolutionary fuel is likely never to be fulfilled. PdVSA hopes to seek congressional approval for its Orimulsion joint ventures proposal in 1993.

4.3 The Future Exploitation of Venezuela's Heavy Oil Deposits

> Where you stand depends on where you sit.
>
> Graham Allison

In his book about the Cuban missile crisis, Graham Allison wrote: 'diverse demands upon each player shape his priorities, perceptions and issues. For large classes of issues – e.g. budgets and procurement decisions – the stance of a particular player *can be predicted with high reliability from information about his seat*' (Allison, 1971, p. 176. Italics mine). These last words, although meant to describe the decision-making process in the top echelons of government, can be conveniently applied to the case of oil policies in different countries. Every oil-producing nation, after all, has an 'oil seat' unique to itself, i.e. an upstream with particular characteristics. Logically, every 'seat' differs from others in terms of the problems, challenges, demands and opportunities which it is likely to pose and, consequently, each one will elicit different policy responses from its owner, *vis-à-vis* such aspects as foreign investment, exploration and development, and so on. This, as we shall see, is especially true in Venezuela's case, because the characteristics of its 'seat' are such that the government, of late, has back-tracked on a number of policy guide-lines which, like the absolute refusal to grant equity interests, had previously been very much taken for granted.

Undoubtedly, the fact that Venezuela is sitting atop staggering quantities of very high sulphur, extremely viscous, and metal-laden heavy crude – and almost nothing else – is the main determinant of Venezuelan oil policy.[26] The country has reconciled itself with the idea that, in the future, it must find a way to tap more of its heavy and extra-heavy crude oils, even if such a move will inevitably entail higher production costs, marketing problems and lower earnings. It is not surprising, then, to find that PdVSA's 1991 production plans foresaw the share of under-22 gravity crude flows growing to 32 per cent (or 1 mb/d) of its total crude production of 3.2 mb/d in 1996; its projections for the end of the 1990s contemplated production of under-22°API crude reaching the 2 mb/d mark, with Orinoco crudes accounting for most of the increment.

However, these plans have what could be a crucial drawback: they were designed under the assumption that the foreign oil companies which have signed letters of intent for the so-called strategic alliances with PdVSA to build deep conversion refineries to upgrade Orinoco crude are actually going to go ahead and build facilities of this kind. Since these refineries are supposed to start receiving 400,000 b/d of Venezuelan extra-heavy oil output by the beginning of the twenty-first century,[27] PdVSA's plans would have to be substantially revised downwards, if not scuttled, should its much vaunted strategic alliances fail to materialize. Fortunately for the company, however, it is not geology alone that determines where international oil companies' investment money goes. Issues like local politics, taxation, production-sharing agreements and royalties, have a great bearing upon an oil company's decision to plough money into an upstream venture. (Indeed, if this were not the case, Venezuela's drive to attract capital to an area characterized by a frustrating geological structure and daunting logistical problems, would be doomed to failure even before it got started.) Unfortunately for the company, the government's decision to cut its investment budget has undermined its financial base significantly, and this has basically put the fate of the Orinoco heavy oil programme at the mercy of international oil companies.

Nevertheless, PdVSA has recognized its geological weaknesses, and has offered equity crude reserves to foreign companies which finance modules to produce and upgrade crude in the Orinoco region. So far, however, all firm offers of equity have been restricted to the Orinoco Belt area; the possibility that equity stakes in other, more amenable, zones of the Venezuelan upstream could be offered has also been raised, but any company wishing to win such a prize would have to make a firm commitment to develop the Orinoco as well. (So far, firm proposals for the granting of equity reserves in 'conventional' areas have been restricted to the Delta Amacuro region, a province where not a single drop of 'conventional' oil has ever been found). This has not pleased PdVSA's prospective partners very much, notwithstanding some of the positive characteristics of the Orinoco area (i.e. production costs of around $3 per barrel, non-existent finding costs, and a virtually unlimited reserve base). It seems that a $4 billion price tag per upgrading facility and a 10–15 year project payback time are still conditions daunting enough to persuade oil companies that there are richer and easier pickings to be had elsewhere (Russia, for instance).[28]

It must be emphasized, of course, that if the strategic alliance plan were to succeed fully, this would have great implications for the world oil market at large, simply because, as one PdVSA executive put it, 'the first 100,000 b/d of reformulated gasoline from bitumen [would] make Venezuela the largest oil province in the world' (*EE*, December 1991, p. 19). However, while this happens, the extremely adverse economics that caused the stillbirth of the Guanipa 100+ and DASM projects are likely to continue to frustrate PdVSA's Orinoco plans. This does not necessarily mean that the world oil market is displaying an abysmal lack of foresight in not tapping this vast source of hydrocarbons. In other words, if Orinoco crude cannot be produced economically at present (and more importantly, anticipated) price and profit levels, this should not be construed as a reflection on the market's ability to absorb the technological risks.[29] A more reasonable conclusion is that the market is signalling that the supply of upgraded Orinoco crude is not economic, because start-up costs remain too high. But today, thanks to the discovery of Orimulsion, PdVSA's fortunes in the Orinoco would appear not to depend exclusively on crude upgrading ventures like those detailed above. This is a massive improvement on the situation confronting the company at, say, the beginning of the 1980s, and it has been entirely brought about by the enterprising skill, resourcefulness and inventiveness of Venezuelan oil men.

PdVSA, acting in accordance with the very reasonable premise that, as yet, Orimulsion appears to be the only economically viable solution to the intractable problem of opening up the Orinoco oil province to large-scale commercial exploitation,[30] earmarked $1.5 billion for expansion of Orimulsion-related activities in its 1991–6 budget. The company wanted to take its Orimulsion production potential to 750,000 b/d by 1996, with an average output of 500,000 b/d (*O&GJ*, 14 January, 1991, p. 36). This implied that, during this period, it would have drilled 1,500 exploratory and development wells (in Cerro Negro, Hamaca and Zuata[31]), built 1,200 km. of oil and gas pipelines, as well as 25 production stations, built a new export terminal, bought four purpose-designed Orimulsion tankers and laid 500 km. of electric power lines (ibid.).

The product dubbed 'liquid coal' has the potential to provide PdVSA with the key to unlock the Orinoco's riches. However, it will have to clear a few major hurdles before it can assume a central role in the world energy balance in the late 1990s and beyond. Green issues constitute the first of these hurdles. As has been seen, Orimulsion is far from perfect from an environmentalist's point of view. Therefore, INTEVEP has continued its research in order to improve the environmental friendliness of Orimulsion, mainly by lowering its sulphur content.[32] However, low sulphur Orimulsion will take between five and six years to reach the market, and unfortunately, this is a long enough period of time to see Bitor's drive to establish Orimulsion as a major contender in the power generation fuel business derailed by utilities refusing to use the product, especially on grounds of its excessive SO_x emissivity.[33] As we have pointed out before, there are a number of alternatives to deal with this problem. First, utilities can install scrubbers, but, the drawbacks of this option have already been enumerated. Secondly, Orimulsion can be burned in conjunction with low sulphur fuel oil, although this is a stop-gap measure which could succeed provided environmental regulation does not increase in severity (as it already has in the USA). Finally, Orimulsion can be turned into gas before undergoing combustion. In the first Orimulsion gasification experiments,[34] Texaco found that an electricity generating plant based on Orimulsion gasification would require lower capital investment than that required for an equivalent coal gasification plant (the notional investment per kW of power produced would be $1,302 for Orimulsion and $1,410 for coal). The Orimulsion gasification plant would also have lower operating costs. Furthermore, Texaco remarked that Orimulsion's high sulphur content would pose no problem for a gasification plant, since the process converts sulphur into hydrogen sulphide gas (H_2S), whose emissions can be removed almost in their entirety by proven methods (WPA, 10 June, 1991, p. 2).[35] Gasification, in other words, could give Orimulsion a viable toehold in the electricity generation sector, even if environmental regulations were to get tougher, provided of course that the cost of the product encourages the installation of more gasification reactors (because at present, there are in the world only about 30 generating plants with 140 gasification reactors, and this number is clearly not enough for an expansion in Orimulsion production like the one desired by PdVSA).

The second hurdle to be overcome is the political opposition to the product which exists within Venezuelan oil circles, where the dispute as to the exact value of Orimulsion for PdVSA is still far from settled. The *Energy Economist*, for instance, implied at one point that the Venezuelan government has given up the possibility of ever selling Orinoco crudes as crude (i.e. a refinery feedstock) and that, in consequence, they would devote their undivided energies to the marketing of Orimulsion.[36] In truth, however, there is nothing to support this view. Celestino Armas, for example, never did warm up to Orimulsion during his tenure at the Venezuelan ministry of energy, and his attitude was said to be shared by other key figures in top Venezuelan oil circles (*PON*, 9 December, 1991, p. 2). This has resulted in a struggle between PdVSA management and the bureaucracy of the ministry of energy, which has sought to cut the funds allocated for Orimulsion expansion, arguing that domestic refining, exploration for light crudes, or even foreign asset acquisition offer better returns on capital (*O&GJ*, 13 January, 1992, p. 18). Opposition to Orimulsion in energy ministry circles can be said to stem partly from traditional organizational values, which assign to the development of oil the highest possible importance, to the detriment of other activities.[37] Perhaps when energy ministry bureaucrats understand that a full-scale development of the Orinoco Belt as an *oil* province is unlikely in the medium term, the threat posed by internal Venezuelan politics to Orimulsion may be lessened. However, the 1992 cuts in PdVSA's budget (which seriously jeopardize the prospects of Orimulsion's expansion), reveal that this notion has yet to sink in.

The threat which Orimulsion poses to heavy fuel oil, however, may generate a very big problem, that is, political flak within OPEC. Granted, PdVSA's avowed objective is to have the fuel compete with coal, but as some have pointed out, 'given the ease of transfer from HFO burning to Orimulsion burning, power companies are not going to switch coal-fired capacity to the new fuel, when the job is more simply done by backing out HFO' (March 1988, p. 16). Thus, if Orimulsion were to take off commercially in a strong way, the other OPEC members would be tempted to include it in Venezuela's production quota.[38] Indeed, the possibility that such a thing might happen looms large in the minds of Venezuelan policymakers, because, as the *Energy Economist* correctly points out, 'PdVSA's strategy depends on Orimulsion's exclusion from OPEC quotas' (*EE*, March 1988, p. 16). In order to prevent this eventuality, the company has gone to great lengths to decouple this potentially revolutionary source of energy from its crude production operations. This task has been simplified somewhat by the fact that both the General Agreement on

Tariffs and Trade (GATT) and OPEC define crude as a liquid at 14° Celsius.

However, even though Orimulsion does not satisfy this admittedly arbitrary definition, PdVSA is taking no chances of it ever being confused with crude. First, as we have seen, it has described it as a competitor for coal, not fuel oil. Secondly, it has priced it accordingly (i.e. at coal parity). Finally, rather than entrust the marketing of the new product to its established integrated oil subsidiaries, PdVSA has formed a new company, Bitor, for this purpose. The new company's name emphasizes the non-crude nature of the commodity it sells – Bitor stands for *Bit*úmenes *Or*inoco. As of now, Orimulsion is still a rather minor player in the boiler fuel market. However, if eventually it has the kind of commercial success many predict for it, it could back out high sulphur fuel oil (and not just coal) as the fuel of choice for many power stations in the world (ibid.). If this were to happen, the Venezuelan claim that Orimulsion competes only with coal would bring little relief to her *confrères* in OPEC, who would be awash in a sea of fuel oil. Under these conditions, it would be logical to expect them to haggle with Venezuela as to how Orimulsion is to be classified. The problem this might pose for the Venezuelans is that their claims that Orimulsion is not a crude or a crude product could be significantly undermined by the fact that it is produced – and will continue to be produced – from reserves which the Venezuelans themselves classify as crude![39]

External political threats to Orimulsion, however, are not restricted to OPEC. Indeed, the possibility that consumer countries might make Orimulsion the target of discriminatory trade practices, should not be discarded. Currently, it is not considered a petroleum product by the customs authorities of the most industrialized nations in the world. But this situation could change quite easily if, say, the refinery rationalization of the EC were to be undermined by key fuel oil users' (like Enel or the Spanish utilities) incremental demand for fuel being met largely by Orimulsion. If this were to result in an over-supply of low value products, the EC could conceivably start treating Orimulsion as a refinery product (it does, after all, go through a manufacturing process), and put duties on it, thus diminishing its commercial attractiveness. However, since Orimulsion supply is a monopolistic affair, it is unlikely that any utility will satisfy its incremental demand entirely with this product (because any availability disruption would have serious consequences for the utility).[40]

What about competition from 'Orimulsion clones' produced in other parts of the world where extra-heavy crude and bitumen accumulations are present? Could they, in time, become another obstacle in its path? Some analysts seem inclined to answer in the affirmative, assuming that, given Orimulsion's apparent technical success, the production of these clones is sure to happen eventually, since 'the emulsification technology can also be applied to Canada's huge tar sands in Alberta, as well as bitumen reserves in Nigeria, Iran and elsewhere'[41] (*EC*, 16 August, 1991). However, the exploitation of tar sands and extra-heavy crudes is still very much a site-specific affair, requiring a conjunction of many factors to create an acceptable level of profitability. Thus, fortunately for PdVSA, it seems that, outside of the Orinoco Oil Belt, geological conditions conducive to the economic production of Orimulsion clones are hard to come by. In fact, geology effectively rules out the possibility of any competitive Orimulsion clone being produced with hydrocarbons coming from deposits located anywhere from near surface depths up to approximately 500 feet. This is because, on the one hand, at depths between 300 and 500 feet, extra-heavy crude is largely lost to exploitation, being too deep to mine and not deep enough for an economical application of enhanced oil recovery methods. On the other hand, accumulations lying near the surface, up to a depth of 300 feet, are exploitable (by means of strip mining with power shovels, draglines and similar equipment), but the complexity of strip mining operations[42] – resulting in very high fixed costs[43] – also makes a non-starter of any project seeking to emulsify bitumen from strip-mineable deposits, such as the whole of the Nigerian tar sands, or parts of the Athabasca tar sands.[44] All of this means that the only deposits that could become the source of raw materials to produce Orimulsion clones are those located below 500 feet, with a high degree of oil saturation, and containing extremely viscous hydrocarbons amenable to recovery by known thermal methods.

Apart from the Orinoco Oil Belt, only the bitumen accumulations of the Athabasca tar sands seem to satisfy these prerequisites. However, most of the heavy oil recovery projects operating in the Albertan tar sands are small in size (from 1,000 to 10,000 b/d of production; see *O&GJ*, 20 April, 1992, p. 78), and unable to offer any real prospect of competition to Venezuelan Orimulsion. The only project big enough to cause Bitor any potential loss of sleep is Imperial Oil's Cold Lake EOR project (80,000 b/d production in 1991, with expansion plans in the wings[45]). However, crude from Cold Lake is of a rather different type than the Orinoco bitumens which constitute Orimulsion. Its gravity (10.2° API), for instance, makes it more akin to very heavy, but still 'conventional', crudes (say, the crudes of the San Joaquín Valley in California), than to Orimulsion material (whose gravity oscillates around 7–8° API). Because of this, Imperial Oil has so far been able to find satisfactory refinery outlets for its Cold Lake crude,[46] to the extent that the company has not even built an upgrading plant – contemplated for the project in its original form.[47] Thus, since Orimulsion can be delivered to the North American market for about $8–9 per barrel (*EC*, 16 August, 1991), it is reasonable to conclude that the Venezuelan product need not fear having an Albertan equivalent for as long as Cold Lake blend (or something similar to it) continues being sold at prices which give the Albertan operators a higher netback than the one they would realize by fielding an Orimulsion substitute and competing with PdVSA's delivered price.[48] In any case, any of Orimulsion's potential competitors would have to overcome a protracted development process, and they would also be afflicted with the Venezuelan fuel's environmental drawbacks.[49] However, the inevitable delays in Orimulsion operations that will follow the government's decision to scale back PdVSA's non-oil operations effectively increase the chance of a competitor for Orimulsion eventually emerging.

Notes

1. Since exploration for petroleum began, upwards of 35,000 oilfields have been discovered (according to the 1991 *Oil&Gas Journal* production report, there are 32,975 producing oilfields in the United States alone).

However, the importance of each of these fields relative to world oil reserves varies enormously. The 37 supergiant fields discovered to date, for instance, contain 51 per cent of world oil reserves. When all the giant and large oilfields are added (this constitutes, at most, 5 per cent of all known fields), they account for 90 to 95 per cent of the conventional oil to be found in the world. In other words, more than 90 per cent of oilfields are insignificant in terms of total world petroleum resources (Riva, 1991, *passim*).

2. The Athabasca tar sands deposit is the world's largest *self-contained* accumulation of hydrocarbons (625 billion barrels). In other words, it is at least four times as large as the largest of all the world's crude fields, Ghawar, in Saudi Arabia (Demaison, 1978, p. 203).
3. According to Martínez (1987, p. 130), the four major producing areas, covering 22,000 sq. km. (40 per cent of the surface area of the belt), account for approximately 84 per cent of the total oil in place in the Orinoco Belt.
4. CVP had shown interest in developing the belt since 1966, but its limited expertise prevented these grandiose plans from being carried out. Then, in 1973, the Venezuelan government approached Henry Kissinger, USA secretary of state, with a proposal for a government-to-government development agreement for the region. The USA denied Venezuela the possibility of such an arrangement on the grounds that 'bilateral deals would set off a bidding war, pitting one consuming nation against the other' (Coronel, 1983, p.43). Later on, the Orinoco province became the jurisdiction of a special section of the energy ministry, which vocally opposed turning the development of this region over to PdVSA. Once the Orinoco project was made a responsibility of PdVSA – after a political struggle with the pro-CVP lobby at the ministry of energy and mines – PdVSA and INTEVEP staff who examined the information on the Orinoco Oil Belt gathered by CVP and the ministry found out that 'the five years of evaluation work done . . . had proved to be an almost complete waste of time' (ibid., p. 115).
5. Following the merger of Corpoven and Meneven in 1986, the Machete area was reassigned to Maraven, while Corpoven remained in charge of the Hamaca area.
6. This certainly does not represent what McGowan (1990, p. 920) says was the Orinoco's share in the total oil investment of PdVSA in these years: 'In the early 1980s . . . hopes were pinned on the development of the Orinoco belt as its rising share of [total PdVSA oil] investment reflected (up from 25 per cent in 1976 to 37 per cent by 1982 [*sic.*])'. McGowan's completely mistaken assertion rests on his identification of PdVSA's investments in the Orinoco with its investments in the Venezuelan eastern region as a whole, which *were* in fact responsible for 25–40 per cent of PdVSA's investments over these years. The lion's share of PdVSA's investments in the East, however, was *not* constituted by expenditures on the Orinoco, but rather by the massive exploration effort intended to find new light and medium crude reserves.
7. Even if Venezuela had succeeded fully in its drive for a larger production

capacity, the way out of the woods into which Pérez's spending spree had led the country would not have been easy to find, as the following lines show. 'The oil industry estimated that, even assuming increases in export prices of 13 to 15 per cent per year . . . and even assuming that the country could develop a production potential of 2.8 mb per day, the country could still develop serious problems of balance of payments and fiscal deficits during the 1990s' (Coronel, 1983, p. 167).

8. According to Coronel (1983, p. 167) sustaining a production level of 2.8 mb/d, would have meant that new areas – especially the offshore and the Orinoco Oil Belt – should have contributed about 1.5 mb/d; of this volume, the Orinoco area alone should have contributed about 1 mb/d.
9. Hence the acronym by which it was known: DASM, which meant Desarrollo Area Sur de Monagas (Development of Southern Area of Monagas).
10. Extra-heavy crude oil upgrading can be accomplished by reducing the amount of carbon relative to the amount of hydrogen in the crude (carbon rejection), by adding hydrogen to the crude (hydrogenation), or through the action of a catalyst. Carbon rejection can be done by either solvent deasphalting or coking. The carbon rejection coking process, though conventional from a technological point of view, has the one drawback of being extremely costly in terms of crude throughput: as much as 30 per cent of the crude run through the coker ends up as coke. See Meyer, 1986, pp. 48–49.
11. PdVSA, it should be noted, was not alone in succumbing to the mirage of eternally high oil prices which would have made projects like DASM feasible. Imperial Oil's (a subsidiary of the Exxon Corporation) Cold Lake project in Canada was even more ambitious. Its stated objective of producing 140,000 b/d of bitumen called for the drilling of 8,000 wells, and the building of a huge steam generating plant. Total investment would have been around $11 billion, or nearly $80,000 per b/d (see Boy de la Tour, Gadon and Lacour, 1986, p. 452).
12. In all probability, PdVSA chose a simpler approach for this second project as an insurance policy, in case the *avant garde* approach failed to come up to expectations.
13. However, the testing of Orimulsion in various power stations has not been without its problems, as this *verbatim* extract from the *Energy Economist* demonstrates:

> In relation to the 'dirtiness' of Orimulsion, BP Bitor recently received a major public relations setback in relation to PowerGen's Richborough power plant test with the fuel. Tar and deposits from the stack settled over a large car park, used by French importers as storage for new vehicles. Naturally enough, with suitable 'economy with the truth' PowerGen's spokespeople told a gullible press that they were 'testing a new fuel called Orimulsion'. Of course, the local TV and newspaper hacks immediately presumed that the incident, in which numerous cars were ruined, was entirely the fault of some backward Venezuelans and their 'disgusting' fuel.

In fact, PowerGen in its immaculate wisdom started up an ancient station in which oil particulate had been deposited over many years in the stack. Vibration loosened these deposits and they crept up the stack at the speed of a snail and slid over the tip and thence onto the finest fruits of the French and German motor industry. It had absolutely nothing to do with Orimulsion, according to Her Majesty's Inspectorate for Pollution. However the ladies and gentlemen of the local press found a suitable scapegoat. PowerGen preened itself on getting away with it. And in BP – co-partner with Bitor in the business – no noise was heard but that of press officers diving beneath their desks. (Poor old Bitor was not exactly in a position to object to PowerGen's 'explanation' for the simple reason that the customer is always right, even when he is totally wrong!) (*EE*, December 1991, p. 20)

14. This is especially true when one compares Orimulsion with what could be considered its closest equivalent, coal slurries. Up to now, coal slurry technology has proved an acute disappointment, since not even the most advanced power plants using coal/water mixes (Tokyo Electric's Yokosuka units and Joban-Kyodo's Nakoso unit) have shown unqualified success. To make matters worse, coal slurry proponents have not been able to overcome the opposition to it by the USA railroads, or to lay at rest the misgivings which environmentalists and water authorities have voiced over the slurry pipelines' high water needs (*EE*, March 1988, p.14).
15. An uncalled for bonus of this procedure is that it lengthens the mean that has to elapse before the wells require a new steam soak, something which significantly improves the product's economics (Reader, 1992, p. 31).
16. A pipeline linking Morichal with an export terminal at Jose is currently being built. The terminal at Jose will have an export capacity of over 50 mt per year, and once it is finished, Punta Cuchillo will be phased out of service (ibid).
17. These deals, involving what amount to sulphur discounts, are strictly *quid pro quo*, as PdVSA's negotiations with Enel (in which the Venezuelans have declared that no sulphur discount will be forthcoming unless the utility installs five gas desulphurization units) show (*WPA*, 10 June, 1991, p. 2).
18. The company's marketing package for its long-term supply contracts includes 'a commercial/technical review for each customer; an engineering feasibility study; mutual price negotiation; a site modification review; supply arrangements and performance monitoring' (*PT*, 4 May, 1990, p. 6).
19. Bitor thinks that its industrial customers would come from the ranks of 'companies prepared to take a long term view of energy use requirements'. This is another way of saying that its customer base will be limited to companies daring enough to tie up their future energy requirements in 10 to 15 year-long supply contracts with a Latin American country, which happens to be a member of OPEC to boot.

20. See for example, Reader: 'Generally speaking, [Orimulsion is] an environmentally friendly product . . .' (1992, p. 31).
21. This high metal content can have its advantages. The Japanese, according to the *Energy Economist* (December 1991, p. 19), use Orimulsion ash as a commercial source of vanadium. On the down side, Orimulsion's nickel content flagrantly exceeds the maximum nickel content limit which Enel, Italy's electric utility and Europe's largest fuel oil user, has set on the fuel oil it purchases (20 parts per million). This is significant because Enel's is the first attempt to regulate nickel content in fuels, allegedly because of the carcinogenic potential of nickelbearing ash. Presumably, all future regulation relating to this matter will take 20 ppm as its starting point, in which case Orimulsion will be hard pressed indeed if it is to comply with such a limit.
22. It also gives Orimulsion the potential of being used in more energy efficient power generation systems, such as combined-cycle power plants; or even as a petrochemical plant feedstock.
23. BP's interest in Orimulsion is long standing. It was a joint technical agreement struck between BP and PdVSA in 1982 which led to the discovery of the product.
24. According to *PON*, 'devising an economic means of shipping Orimulsion has been a leading factor hampering development of the Far East market [for the product]'(5 December, 1991, p.1). To remedy this, PdVSA and Mitsubishi have discussed the possibility of constructing a trans-Costa Rican pipeline, to make the transportation of the product easier. The delivery procedure would be similar to the one employed to transport crude across the Panama canal: an Orimulsion tanker from Venezuela would berth in the east coast of Costa Rica, where it would unload the product, which would then be transported cross country, to another tanker waiting at the far end of the pipeline. So far, this project has not gone beyond a mere conceptual study, however.
25. The German utility MVS will begin testing Orimulsion in its 377 MW Marbach power station from mid-1992. Enel has signed a letter of intent to buy 1 mt of Orimulsion annually for five years.
26. As one PdVSA executive metaphorically put it, 'Venezuela's problem child, heavy oil, must be adopted before more attractive offspring are considered' (*WPA*, 27 January, 1992, p. 17).
27. The rest of the country's output would be divided thus: 650,000 b/d to refineries in Venezuela and Curaçao, 450,000 to PdVSA's overseas refineries, and 500,000 to third party sales (*PIW*, 4 November, 1991, p. 8). Marketing to third parties could also have undertones of vertical integration, since it has been proposed that it be conducted through preferentially supplied deals which would allow foreign refiners to upgrade their plants to handle the crudes.
28. As Roel Murris, former director of exploration at Shell Internationale Petroleum Maatschapij, put it: 'There are concerns in parts of Africa and South America . . . that the potential availability of new areas in the CIS . . . will cut the amount of money available for investment. Certainly, the amount of money available for exploration is not

infinite, so, on the face of it, *those concerns may be justified because as in any other business, capital and expertise will be attracted by the best opportunities* (*Shell World*, April 1992, p. 25. Italics mine).

29. A conclusion much beloved by Venezuelan oil men. See Juan Chacín's statements in *PdVSA Contact*, July–August 1988.
30. When PdVSA's plans for commercial development of the Orinoco were scaled back in the mid-1980s, the pilot installations originally designed to evaluate production systems which could have been used for the full-scale operations were left as the only production infrastructure in the belt. For years, these plants produced small amounts of extra-heavy crude, which had to be reconstituted with naphtha and/or blended with lighter streams before being refined or sold. The economics of this operation were always quite marginal.
31. In other words, if all had gone according to plan, Cerro Negro would no longer have been the sole producing area for Orimulsion. PdVSA hoped to link the Zuata and Hamaca production areas with processing centres located at Oficina, with the Orimulsion being pumped through a common pipeline, from Oficina and Morichal, to a new export terminal on the Caribbean (Reader, 1992, p. 30)
32. It should be said in passing that Orimulsion's production process itself also poses some environmental drawbacks. The steam soak stimulation technique demands considerable amounts of fresh water on the production site. Even if the production were to multiply more than seven-fold, as PdVSA has planned, water needs for the product could still be comfortably met, given that the Orinoco is one of the great rivers of the world. Waste water disposal, however, could be problematic, since the increasingly vocal environmental groups (in Venezuela and elsewhere) would presumably be less than thrilled at the sight of millions of m^3 of oily water and sludge being dumped into the river every day.

    An illustration of the great water needs of a heavy oil enhanced recovery facility is provided by Imperial Oil's Cold Lake project. Imperial has been warned by the Albertan government that it might face closure if the water supplies in the region drop from the present level. Imperial uses 3.65 milion m^3 per year of water, and has been blamed by local residents for the reduction of Cold Lake's water level to its lowest point in 30 years (*PON*, 17 December, 1991, p.3).
33. A supply deal with Florida Power and Light, which had been hailed as a breakthrough for Orimulsion in the US market, failed on these exact grounds. The utility concluded that 'despite the pricing of Orimulsion on a level competitive with coal, equipment and cleanup costs eliminate[d] its competitive stance' and that, therefore, it would stick to gas as its fuel of choice for its new facilities (*IGR*, 20 March, 1992, p. 18). This was a major blow for Orimulsion's commercial prospects in North America.

    PowerGen, in the UK, has seen its plans to burn Orimulsion in its Pembroke (Wales) power plant come under severe criticism on account of SOx emissivity. Environmental groups have argued that introducing Orimulsion burning in the station will increase the emissions of this pollutant six-fold. That statement, however, omits to mention that the station's emissions have been abnormally low because it has been operating at 10 per cent of its capacity.

34. Texaco gasified Orimulsion for the first time in 1989, at its Montebello research laboratory in California.
35. In June 1992, Texaco and Bitor America signed an 'agreement to develop integrated gasification combined cycle power projects using Texaco's gasification technology and Venezuelan Orimulsion as feedstock'. Other potential partners in this project include General Electric and Teco Power Service (*WPA*, 19 October, 1992, p.4). The first results of this agreement have been talks on the possibility of constructing an Orimulsion gasification plant in Puerto Rico, to provide around 250 MW of power.
36. Witness this quote as an example: '[PdVSA has come] to the very clever conclusion that it would be best to stop thinking of it [Orinoco crude] as crude at all, add water at considerably lower cost than any refinery operation, and sell it as a new power generation fuel . . . [which] can beat the socks off both coal and HFO and find itself a formidable market share' (*EE*, March 1988, p. 19).
37. PEMEX, for instance, has always considered the exploration and exploitation of natural gas to be ancillary to exploration and exploitation of crude oil, notwithstanding the great demand which exists in Mexico for natural gas (see *WGI*, March 1992, pp. 5–7).
38. When PdVSA began to market Orimulsion, Niering wrote: 'the advent of Orimulsion raises an interesting – and potentially contentious – question for OPEC, which at some point, presumably depending on the extent of the new product's success, will be asked to decide whether it should be included in Venezuela's quota' (1989, p. 220).
39. Until 1992, all Orimulsion production has come from Orinoco deposits which the company classified as 'easily recoverable by conventional means', and have therefore been included in Venezuela's crude reserves tallies.
40. The fact that PdVSA is a monopolistic supplier for Orimulsion may also discourage potential industrial customers from entering into the long-term contracts desired by the Venezuelans.
41. Demaison, 1978, gives a description of world bitumen and extra-heavy crude deposits. For a description of Nigerian tar sand resources see Ako *et. al.*, 1983. On Canadian tar sands, see Dorsey, 1983; Niering, 1980; Wennekers *et. al.*, 1979; *O&GJ*, 11 February, 1980, pp.36–38; *Oilweek*, 24 March, 1980, pp. 16–28.
42. The logistics of a bitumen strip mining operation (like Syncrude or Suncor, operating in Alberta) are as follows. First, the soil overburden covering the deposit is set aside, to be used later for the reclamation of the site. The oil bearing rock is mined, and sent to a processing facility where it is physically separated from the rock, usually by a hot water process or by the use of a solvent. For all its apparent simplicity, however, this operation involves much more than digging and boiling rocks. See *O&GJ*, 11 February, 1980, p.36.

43. Good (1985, p. 8) wrote that variable expenses account only for about 15–20 per cent of the production costs for Canada's Syncrude project. Syncrude's production costs in 1983 and 1984 were 195 and 208 million Canadian dollars, respectively, even though 1984 output represented a 23 per cent decline relative to 1983 production levels.
44. McGowan (1990, p. 923) says that producing an Orimulsion imitation from mined Canadian oil sands would cost as much as $9–12 per barrel. Strip mining has additional disadvantages because of its extremely deleterious environmental impact, and the major problems it poses regarding the continuous handling of vast quantities of materials (for instance, in order to mine 200,000 cubic yards/day of tar sands – a reasonable volume for a 100,000 b/d processing plant – two to three times this volume of material has to be stripped and, eventually, replaced). In fact, as Wennekers *et. al.* (1979, p. 292) have pointed out 'tar sands mining [has only one] strategic advantage . . . [namely that] in contrast to heavy oil production [by means of EOR methods], tar sands open pit mines produce for a design life of at least 25 years at a steady "base flow" production rate, whereas in heavy oil production, each good oil well decays exponentially within 8–12 years.'
45. An increase to about 100,000 b/d by 1992 is currently under consideration. Increased productive capacity was added in 1988 at Cold Lake, but given a weak heavy oil demand, it was mothballed (*O&GJ*, 28 October, 1991, p. 28). Water shortage problems, however, may kill the whole scheme.
46. Crude from Cold Lake is blended with 68.8–70.3° API condensate, resulting in an export stream called Cold Lake blend. The gravity of this blend is 22.6° API (blended Orinoco crude from the Guanipa 100+ project, by contrast, would have had a maximum gravity of 16° API).
47. According to some sources, Imperial is considering the construction of a heavy oil upgrader, tied to its heavy oil operations in Cold Lake. The project, with a $3 billion cost, is seen as an alternative to the proposed $4.5 billion OSLO oil sands project, located in the Fort McMurray region, and which recent studies have shown would not be economic at current oil prices until after 2000 (*O&GJ*, 28 October, 1991, p. 28).
48. This will be even more difficult in the near future (1993), since PdVSA expects to reduce Orimulsion's production costs by 25 to 35 per cent by using jet kerosene as surfactant in the primary phase of production (*PON*, 5 December, 1991, p. 1).
49. This is a crucial point, because the natural markets for any viable Orimulsion clones would be California and the US East Coast (due to their proximity to the producing areas). However, these places are under the sway of the most restrictive environmental laws to be found anywhere. This means that Cold Lake's potential location advantage *vis-à-vis* Orimulsion in respect to the US East Coast market is of no consequence, because Cold Lake bitumen could not be marketed there. Therefore, exporting Cold Lake bitumen in the form of an Orimulsion clone would necessitate building a seaside export terminal, having dedicated tankers and so on.

5 NATURAL GAS IN VENEZUELA

5.1 Natural Gas Deposits

Venezuela has the seventh largest proved reserves of natural gas in the world[1] (3.43 trillion m^3, representing 2.6 per cent of world reserves). The main gas deposits are found in the Oriental basin, mostly in the Anaco area. This basin contains 66 per cent of the country's total gas reserves, with the Lake Maracaibo area accounting for a further 23 per cent, and the recently discovered offshore fields to the north of the Pariá peninsula making up the remaining 10 per cent (CEPET, 1989, v. 1, p. 361). As Figure 5.1 shows, Venezuelan gas reserves have experienced a period of marked growth from the mid-1980s onwards. The single largest increase in reserves came in 1985–6, and it was due to the great quantities of gas found in association with crude in the new oilfields of the Oriental basin (PdVSA, 1986, p. 15).

The chemical composition of the natural gas found in different areas of Venezuela is quite variable, (see Table 5.1). The

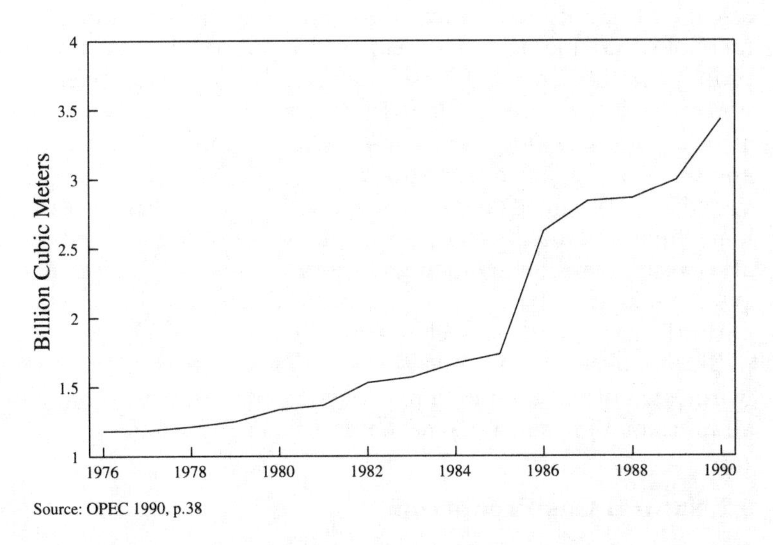

Figure 5.1: Evolution of Venezuelan Gas Reserves. 1976–1990.

Table 5.1: Composition of Natural Gas in Different Areas of Venezuela (Percentage of Weight).

Component	*West (A)*	*Guárico (N)*	*East (N)*	*East (A)*	*Offshore (N)*
Methane	73.1	83.5	76.9	75.1	90.5
Ethane	11.0	0.6	5.8	8.0	5.0
Propane	6.0	0.1	2.5	4.6	2.2
I-Butane	1.1	-	0.5	0.9	0.4
N-Butane	1.9	0.1	0.6	1.1	0.7
I-Pentane	0.6	-	0.3	0.3	0.3
N-Pentane	0.5	-	0.2	0.33	0.2
Hexane	0.5	-	0.2	0.2	0.2
Heptane	0.4	-	0.4	0.2	0.2
Carbon dioxide	4.4	15.6	12.5	9.2	0.2
Nitrogen	0.5	0.1	0.1	0.1	0.1
Specific gravity	0.8	0.7	0.8	0.8	0.6
Calorific content (Btu/cf)	1,273	857	1,033	1,126	1,136

A: Associated gas
N: Non-associated gas
Source: CEPET, 1989, v. I, p. 360.

associated gas of Lake Maracaibo and Anaco has greater commercial value than that of other areas, because it is quite rich in propane and butane.[2] It is also rich in ethane, a valuable fraction used as a chemical plant feedstock. In contrast, the non-associated gas found in the Guárico basin (Yucal-Placer area) and the Oriental basin (Lechozo and other fields) has the significant disadvantage of containing a great amount of carbon dioxide, which not only lowers the calorific value of the gas, but can also cause 'corrosion in pipelines in combination with water, and excessive heat to be liberated in hydro-desulphurisation when natural gas is used as a chemicals feedstock' (Peebles, 1992, p. 3). Furthermore, the Yucal-Placer gas is quite sour (in other words, it contains too much hydrogen sulphide). Natural gas in all areas of Venezuela is reasonably free from nitrogen.[3]

5.2 Natural Gas Production

In Venezuela – as in many other parts of the world where the appropriate infrastructure to take gas to a market did not exist – natural gas was initially considered a nuisance, an undesirable by-product of oil production. As a consequence, up to the year 1945, almost all Venezuela's natural gas production literally went up in flames, because flaring it off into the atmosphere was seen as the cheapest alternative for its disposal. However, from 1946 onwards, the companies operating in Venezuela found a use for natural gas: reinjection, to help maintain the flow of oil in wells where natural pressure had fallen (in that year, 27.5 million m^3 of gas were flared, out of a total production of 31 million m^3; gas used for injection accounted for the balance). Gradually, as the use of gas injection became widespread, and the appreciation of the versatility and importance of natural gas both as a source of energy and a petrochemical feedstock grew, the wasteful practice of flaring has been reduced. Table 5.2 shows the evolution of natural gas use in Venezuela since 1973.

Table 5.2: Production and Use of Natural Gas in Venezuela (Millions of Cubic Metres per Year).

Year	*Production*	*Injection*	*Fuel**	*Manufacturing*	*Commercial*	*Flared*
1973	49.433	21.790	5.791	1.721	5.512	14.619
1974	46.425	23.085	5.698	1.846	5.934	9.862
1975	38.008	20,968	4.912	1.823	6.016	4.289
1976	37.135	20.500	5.166	1.930	6.492	3.047
1977	37.512	19.856	5.392	2.104	7.332	2.828
1979	34.842	17.821	5.345	1.889	7.496	2.291
1978	36.943	18.317	5.692	1.916	8.960	2.328
1980	35.499	16.534	5.562	1.876	9.242	2.235
1981	34.681	16.119	5.462	1.756	9.424	1.920
1982	33.526	13.860	5.966	1.725	9.917	1.788
1983	31.766	12.866	5.093	1.525	10.547	1.735
1984	32.574	12.030	5.191	1.488	12.109	1.756
1985	32.996	12.428	5.043	1.625	12.283	1.617
1986	36.275	12.040	5.780	2.386	13.294	2.775
1987+	36.233	11.382	5.431	2.362	13.411	3.647
	Production	*Injection*	*Manufacturing, commercial sales, fuel**			*Flared*
1988	37.960	13.870	21.535			2.555
1989	37.960	12.775	21.535			3.650
1990	40.515	13.140	24.090			3.285

* Fuel for oil industry applications.

+ Last year for which data on each variable are available on a non-aggregated basis.

Sources: CEPET, 1989, v. I, p. 363; PdVSA, 1976–90.

Even though the use of natural gas by the private sector in Venezuela grew substantially during the 1980s, the oil industry remains the major consumer of gas in the country: oil well injection, for instance, still accounts for 32 per cent of production. The low level of gas use in households and industry is a result of limitations on the side of the distribution infrastructure: the gas distribution grid for light industrial uses and domestic consumption is quite small – on account of geographical considerations – and embraces only the cities of Caracas, Puerto La Cruz and Valencia (*IGR*, 12 December, 1991, p. 4). Even so, a 30 per cent increase in direct burn domestic demand over the period 1991–7 is expected (ibid.). Gas consumption may also get a further boost if plans to introduce compressed natural gas as a transportation fuel go ahead.[4] Corpoven expects that 'some 60,000 CNG-fuelled vehicles will be on the road by 1994' (*WGI*, December 1990, p. 17). This forecast seems a trifle over-optimistic, but it clearly reveals the intention of the Venezuelan government to continue fostering gas consumption. This is quite understandable, because throughout the period 1980–91, the Venezuelan government reaped considerable benefits from its strategy of substituting domestic consumption of liquid hydrocarbons with consumption of natural gas. Since natural gas absorbed all the growth in the domestic demand for energy, domestic oil use was kept flat at around 350,000 b/d, a boon for an oil exporter with a dwindling conventional resource base. Domestic gas consumption has freed considerable volumes of liquid fuels for export; in 1990, for instance, natural gas use in Venezuela was equivalent to a consumption of 204,000 b/d, of oil (*O&GJ*, 26 August, 1991, p. 37).

The gas-dedicated infrastructure in Venezuela, although not as big as that dedicated to oil, is none the less quite impressive. The National Trunk Gas Pipeline Network (Red Nacional de Gasoductos Troncales) has a total extension of 4,750 km. To move the gas through this network, PdVSA has 490 compression units (distributed among 152 plants) with a total installed power of 500,000 hp. A major addition to the Venezuelan gas network is the NURGAS network (acronym for Nueva Red de Gasoductos, or New Gas Pipeline Network), which began operations in 1990. This 802 km. pipeline system, with the capacity to transport 28 mcf of gas per day, was built in order to satisfy the growing demand for gas from the central and western areas of the country. When fully completed, it will link the Anaco producing area with Río Seco, in Falcón state.

5.3 Future Developments for Natural Gas

Up until very recently, associated gas accounted for 99 per cent of the total Venezuelan production.[5] Thus, through time, gas production in the country has been a function of crude oil production, which may fluctuate for a great variety of reasons. Nowadays, with natural gas playing an increasingly important role in the supply of energy to the internal market, the prospect of gas availability being curtailed by some unforeseen development in the oil market has become less palatable to the government. Therefore, PdVSA has been ordered to step up its exploration and production efforts for non-associated gas, with the aim of increasing the flexibility of its gas output (*WGI*, December 1990, p. 17). Also, in order to meet the surge in gas consumption expected for the period 1991–6, PdVSA has announced that it will expand its gas related infrastructure considerably: the high pressure pipeline network will be enlarged by 1,250 km. at a cost of $400 million; 28 new gas compression units will be built, in order to add 765,000 hp of compression capacity to the system, at a cost of $1.1 billion; and 138,000 b/d of LPG processing capacity will be added, at a cost of $180 million. The pipeline expansion programme, whose implementation will be wholly left to Corpoven, consists of three major projects. The first project will link the Anaco fields with Puerto Ordaz, a centre for steel and aluminium manufacture on the Orinoco river. The second will link the Morón complex with the two refineries of Amuay and Cardón. The third project involves linking Anaco with the Jose cryogenic plant and petrochemical site (*EE*, December 1991, p. 20). The objective of all these investments in infrastructure is to take gas production up to a level of 6.14 billion cf/d by the end of the 1990s.

This activity seems to be justified by the forecast growth of the Venezuelan internal market for gas. According to *World Gas Intelligence*, 'Venezuelan internal gas consumption is projected to exceed 23-Bcm/year by the end of the century, growing by an annual 6.8% to meet increased demand for fuel and feedstock. Natural gas is expected to rank just behind hydroelectric power and ahead of liquid hydrocarbons in meeting anticipated energy demand of 1.2 mb/d of oil equivalent in the year 2000' (December 1990, p. 17). However, Venezuela's petrochemical sector will also soak up a great deal of the incremental supplies of natural gas. PdVSA's subsidiary Pequiven will expand olefin output at the El Tablazo complex by 400,000 tonnes per year and ammonia/urea production at Morón by 360,000 tonnes. At the Jose complex, plans contemplate the production of 620,000 tonnes of olefins, 500,000 of MTBE, 1.45 mt of methanol and 500,000 tonnes of ammonia (*WGI*, December 1990, p. 17). Thus, Venezuelan petrochemical production is expected to increase from 2.2 mt in 1990 to 12.5 mt in 1995, effectively transforming the country's status from a net buyer to a major exporter of petrochemical products (8.5 mt per year).[6]

Local distribution of gas is an area whose future developments may prove to be of great interest to investors within and outside of Venezuela. At present, PdVSA is vertically integrated from wellhead to burner tip: Lagoven, Maraven and Corpoven control the production, gathering and processing of natural gas, and Corpoven subsidiary Cevegas controls the distribution. However, a scheme whereby distribution and marketing of natural gas would become a responsibility of municipal authorities – which would be free to invite domestic and foreign private capital participation – is currently under discussion. Of course, as of the moment, the whole scheme does not have a lot of substance, but the mere fact that it has been proposed could prove to be the herald of a major departure from established practice by PdVSA (*WGI*, December 1990, p. 17).

PdVSA has also proposed a scheme which could see Venezuela becoming the supplier of up to a third of Colombia's future gas requirements. The plans contemplate exporting up to 200 mcf/d of gas to Colombia under a 30 year contract, through a pipeline connecting the Venezuelan and Colombian distribution systems. The proposed pipeline would measure 230 km. in length (of which 170 km. would be in Venezuelan territory and 60 km. in Colombian territory). The Venezuelan portion of the pipeline could cost about $100 million, while the

Colombian section would cost $38 million. The interconnection of this pipeline with the Colombian distribution system could be in the north of the country, or near the site of the Barranca Bermeja refinery (*POW*, 4 November, 1992, p.4). The target start-up date for the project would be 1993, with the first volumes of gas coming from Venezuela's western gasfields in 1995. Later on, the supply source would switch to the large eastern fields. There are still plenty of obstacles left in the path of this venture, however. Financing and volume issues still have to be worked out, and there seems to be a major disagreement over prices between the two parties.[7] And, last but not least, is the fact that the Venezuelan congress has to give its approval for the project to go ahead; even if the economics of the undertaking were good, this approval cannot be taken for granted because, traditionally, there has been no great love lost between the two countries (and even less between the politicians of both countries).[8]

PdVSA's most ambitious gas project for the future, however, is the Cristóbal Colón liquid natural gas export project, which involves the development of the offshore gas reserves discovered by Lagoven in the Carúpano basin in 1978 (Mejillones, Patao, Río Caribe and Dragón). Cristóbal Colón, which would be the largest undertaking ever in the history of the Venezuelan oil industry if it were to go ahead in its original form, seems destined to become a landmark in the annals of PdVSA for reasons other than the magnitude of the investment required to bring it to fruition. This is because the project might give foreign oil companies equity participation in the Venezuelan petroleum sector for the first time since the oil nationalization took place in 1976. This radical break with the dogma which guided oil policy during the period 1976–90 (refusing to grant any type of ownership of oil resources to foreign capital) is the result of the recognition by PdVSA and the government of the simple truth that, without the introduction of foreign capital, technology, and expertise, a project of this size and complexity would defeat any attempt on the part of the nationalized industry to tackle it. Therefore, since its very inception, the project has envisaged PdVSA subsidiary Lagoven as holding only 30–40 per cent of the total equity of the project, with the remaining interest owned by a combine of international oil companies (whose contribution to the deal would consist of providing financing and technology for it, and ensuring the access of the Venezuelan LNG to markets in the USA, and possibly, Europe).

PdVSA received bids for participation in the project from BP, Exxon, Mitsubishi, Shell, Texaco and CFP Total. In July 1990, the company chose Shell, Mitsubishi and Total as its partners; Total later backed out and was replaced by Exxon (*O&GJ*, 18 March, 1991, p. 135). According to PdVSA's master plan, Shell, with a 31 per cent share, would be responsible for the drilling and processing technology; Exxon, with a 29 per cent share, would handle the marketing of the gas in the USA; and Mitsubishi, with an 8 per cent share, would provide the venture with its transport expertise. Originally, construction was to have begun in 1992, with the first exports starting in 1996. Estimated revenue over the 20-year lifespan of the project was put at $500 million per year, with a production capacity of 70 mcf/d. Finally, in what amounted to a major departure from established practice in Venezuela, the government established that the owner of the gas would be the joint venture company, and *not* the state (*PIW*, 29 October, 1990, pp. 1–2).

A preliminary development agreement was struck between PdVSA and its three partners in March 1991. This contract called for the partners to conduct studies of the project's economic feasibility, the results of which – once negotiation of fiscal terms between the partners was satisfactorily completed – would serve as a stepping stone for the establishment of a joint venture company. The joint venture agreement, subsequently, would be turned over to the Venezuelan Congress for approval. If this approval was received, the partners would complete the economic feasibility studies and reach a decision on whether to proceed with the project (*O&GJ*, 18 March, 1991, p. 135). However, the start-up date for the project has been displaced to at least 1998 (the original intended date was 1996). This seems to be quite a logical decision. If the 1996 deadline had been maintained, PdVSA and its partners would have had only four years in which to drill 55 offshore development wells in about 394 feet of water, down to pay depths averaging 7,500–8,000 feet; install anything from four to eight production platforms; lay nearly 80 km. of undersea and overland pipelines; build a processing/liquefaction/tanker port complex on the Venezuelan mainland (in Mapire Bay, between the towns of Guiriá and Macuro on the Pariá peninsula); and, last but not least, commission three new LNG tankers (*PIW*, 29 October, 1990, p.1; *O&GJ*, 14 January, 1991, p. 36).[9]

However, even if, as is apparent, the project is running behind schedule, it undoubtedly has had easy sailing so far, compared to the political squalls which an earlier Venezuelan effort to set up an LNG capability in the country had to face.[10] The delays have been due, among other things, to the time it took to reach an accord as to the way the whole project would be taxed by the government; in the end, after much fencing between the Venezuelan bureaucracy and PdVSA, the tax rate on joint ventures has been set at 30 per cent of earnings, a great improvement when compared to the 67.7 per cent oil industry corporate tax the partners would have had to pay before the amendments to the Venezuelan income tax laws (*PIW*, 5 August, 1991, p. 5).[11] Also, Shell's insistence on restrictions to protect its LNG technology from being pirated by partners Exxon and Mitsubishi (major competitors of the Dutch/English giant in the international gas market) has prevented the agreement from crystallizing. Shell, for instance, objected to the proposed inclusion of an Exxon official on the project's technical committee, and it has not been appeased by an Exxon offer to include a confidentiality clause in the agreement which would keep it from using similar technology for a 10-year period after the start-up date of Cristóbal Colón (*PON*, 25 August, 1992, p.1). But the delays have not made the project grind to a standstill; the 3-D seismic survey of the gas bearing structures offshore has been completed, and the gas reserves in the area have been tentatively upgraded to 8tcf (compared with original estimates of 6tcf). This has raised the possibility of expanding the design capacity of the plant to 900,000 mcf/d, to 10–15 per cent of current transport costs over the lifetime of the projects (*PON*, 25 August, 1992, p.4).

As we noted before, if Lagoven and its partners decide to go ahead with the project, the Venezuelan congress would have to approve it. It seems likely that this formality will be completed to the satisfaction of the partners, because there is no great opposition to Cristóbal Colón in Congress, a situation quite unlike the hostile political environment which capsized the 1970 Caldera LNG scheme (although this is not to say that there will not be any minor legal inconveniences in this process of parliamentary revision and approval). However, congressional opposition is not the real enemy of the LNG scheme this time around.[12] 'On the contrary [says the *International Gas Report*] any uncertainty surrounding the project relates to the North American gas market, which seems increasingly reluctant to think long term' (*IGR*, 12 December, 1991, p. 3). According to this source, 'it would be supremely ironic if, after 16 years of internal hostility to foreign partnership, it was the external market situation that let the Venezuelans down' (ibid.). Perhaps this is true, but irony or no irony, the gas bubble which has plagued the North American gas market since the early 1980s has given no sign of abating in earnest,[13] and this fact will make the international oil companies very wary of ploughing their funds into potential white elephants.[14] The experience of the Cove Point terminal in Maryland – which has been in mothballs for years – provides Shell with a poignant reminder of an LNG project gone awry, while Exxon must still remember the technical problems and cost overruns which turned the Libyan Marsa al-Brega LNG plant into a complete disaster.[15] Of late, the companies have turned their sights towards Europe as 'an alternative to exporting the planned 4.4 million ton annual LNG output to the US', but the chances that Cristóbal Colón will have sound economics delivering gas to Europe are slim indeed (*PIW*, 20 April 1992, p. 6).

Complicating matters even more is the fact that, certainly up to the year 2000, global LNG economics will still be very much linked to the comings and goings of the international oil price level in the foreseeable future. The volatility which this can transmit to LNG contracts (since oil prices will still be in thrall of events in the unstable Middle East and the potentially explosive region encompassed by the former USSR) would certainly not be a propitious environment for the high capital investment required to get a project like Cristóbal Colón under way.

Nevertheless, even if the project never gets off the ground (and this writer is among those thinking that this will be the case), it might still serve a useful purpose; namely, that of indicator of the earnestness of Venezuela's about-face regarding foreign investment in its oil and gas sector. As *World Petroleum Argus* has pointed out, the rapid approval of the project by the Venezuelan Congress is seen by foreign companies as a test case of how readily (and willingly) Congress would approve other joint ventures between PdVSA and foreign oil companies, including the strategic alliances for the exploitation of the Orinoco Oil Belt (*WPA*, 27 January 1992, p. 19).

Notes

1. After Russia, Iran, the UAE, Saudi Arabia, the USA, and Qatar (OPEC, 1990, p. 38).
2. These fractions are used in the manufacture of Liquid Petroleum Gas (LPG), a product which commands relatively higher prices than dry gas.
3. Nitrogen is a nuisance because, although it is neither corrosive nor toxic, 'it lowers the calorific value of the gas and increases transportation costs, since larger pipelines are needed to carry the same amount of useful energy' (Peebles, 1992, p. 4).
4. In 1989, the first experimental outlets for the supply of CNG to automobiles were constructed.
5. 88 per cent of the natural gas in Venezuela is found in association with oil.
6. Foreign investors will foot the greater part of the bill for all these projects.
7. Venezuela favours a price of $5 per million Btus, while Colombia would like to see a price of $1 per million Btus. According to *PON* (4 November, 1992, p.4), an external advisor estimated that $2.50 per million Btus price was a fair price. However, neither country has indicated whether it will accept this price.
8. Relations between the two countries are still tense, due to the issue of boundary demarcation in the Venezuelan Gulf. For instance, when foreign minister Humberto Calderón Berti took up his post in March 1992, he declared that Venezuela 'would adopt a more rigid position over its negotiations with Colombia over the . . . Gulf of Venezuela territorial dispute' (*PON*, 12 March, 1992, p. 3)
9. The costs of the various parts of the project are estimated as follows: drilling of the wells and installation of platforms, $1 billion; production plant and terminal, $1.3 billion; LNG tankers, $300 million apiece (*PIW*, 29 October, 1990, p. 1).
10. Venezuelan president Rafael Caldera first came up with the idea of setting up LNG capacity in Venezuela in order to export gas from the fields in the Maracaibo and Anaco areas. This 1970 plan involved the construction of two liquefaction plants and the purchasing of a new, Venezuelan controlled, LNG tanker fleet – seven ships strong – for transporting the gas to the US East Coast. The oil ministry thought that the plants could be ready by mid-1975, and producing income of

about $200 million a year by 1976. The LNG project was seen as a vital source of new income for Venezuela, one which perhaps would make up for the loss of income that would result from the concession areas going into productive decline (a decline which, as some pointed out at the time, would also have cut the volume of gas available for LNG production). (*PIW*, 4 October, 1971, p.2.)

In order to make the financing of the project easier, Caldera originally wanted to work out joint ventures between CVP and the original two proponents of the idea: Creole Oil and The Philadelphia Gas Works. The political mood in the country led him to propose to keep the system wholly in Venezuelan hands, restricting foreign equity to the tanker fleet. Governmental borrowing and domestic funds would have been used to finance the investment.

Caldera's bill proposing the LNG scheme came under savage attack in the Venezuelan congress, especially from the former oil minister Pérez Alfonzo. Congress made so many alterations to the bill that it emerged, in August 1971, in the completely unrecognizable form of a gas nationalization decree(!): the 'Law Reserving the Natural-Gas Industry to the State'. This law asserted state ownership over all associated and non-associated gas, and restricted gas exports to associated gas, thus leaving the fields in eastern Venezuela for the supply of the domestic market (Tugwell, 1975, pp. 127–9). In the end, running the gauntlet of congressional approval so maimed Caldera's plan that it was eventually (and ignominiously) shelved.

11. However, the companies have been reluctant to commit themselves definitely until they have been 'guaranteed a legal and tax framework that can't be tampered with arbitrarily by the Venezuelan government'. (*PON*, 30 June, 1992, p. 2).
12. In the wake of PdVSA's 1991 budget cuts, the fact that its partners in the scheme will be the ones putting up most of the money required is probably the only thing that has kept it from foundering.
13. Although as a result of the records which both spot cash and futures prices set in May of 1992, many analysts are now optimistic about the long-term demand and price outlook, as the following extract proves: '1992 may be the first year in over a decade that demand exceeds deliverability . . . The strong action of gas prices is causing us to think the unthinkable – that the gas bubble may indeed finally be over' (*EC*, 29 May, 1992). Of course, analysts have said this before, and been woefully short of the mark.
14. *PIW* reported in its 20 April, 1992 issue (p. 6) that 'the actual go-ahead for the [Cristóbal Colón] project probably depends on an upturn on depressed US prices'. The c.i.f. price for Cristóbal Colón gas needed for the project to achieve the desired rates of return would seem to price the gas out of the US market. According to *PON* (25 August, 1992, p.4) 'for the project to succeed, its LNG output must be sold for at least $2–3/mcf, a far cry from [what] US spot prices [have been in the 1990s].
15. The delays in the start up of the Cristóbal Colón project have already raised its expected cost by a half billion dollars (*PON*, 30 June, 1992, p. 2).

6 THE VENEZUELAN DOWNSTREAM SECTOR

6.1 The Venezuelan Refining System

The decree which, by virtue of the nationalization of the assets of all the foreign companies operating in Venezuela, gave birth to PdVSA, created overnight one of the biggest refining companies in the world. Unfortunately for the new company, however, the quality of its inheritance was not commensurate to its size (12 refineries with 1.449 mb/d of distillation capacity; see Table 6.1). On the one hand, seven of the Venezuelan refineries were small and obsolete; since they accounted for only about 15 per cent of the total refining capacity in the country, they were strong candidates for closure. On the other hand, the system was completely inadequate to cope with the necessities of the world market for refined products, as well as those of the Venezuelan internal market. This was because it was geared to satisfy a demand pattern which was very different from the one which had developed in the petroleum products market from the beginning of the 1970s.[1]

Table 6.1: The Venezuelan Refining System. 1976. Capacity in Barrels per Day.

	Atmospheric distillation	*Vacuum distillation*	*Catalytic cracking*	*Reforming*	*Desulphu-rization*	*Alkyl-lation*
Amuay	630,000	350,000	-	10,000	230,000	-
Cardón	328,000	80,000	57,600	-	67,600	4,370
San Lorenzo	26,900	-	-	-	-	-
Puerto La Cruz	160,000	-	12,000	-	-	2,000
El Palito	105,000	-	-	7,500	-	-
Bajo Grande	45,000	15,000	-	-	-	-
El Chaure	40,000	-	-	-	-	-
El Toreño	5,400	-	-	-	-	-
San Roque	5,200	1,500	-	-	-	-
Morón	20,000	15,000	-	-	-	-
Caripito	74,200	-	-	-	-	-
Tucupita	10,000	-	-	1,400	15,000	-
TOTAL	1,449,700	461,500	69,600	18,900	312,600	6,370

Source: *O&GJ*, 25 August, 1980, p.69.

Before 1976, refining in Venezuela had always been oriented towards the production of the greatest possible volumes of residual fuel oil (the yield of this product historically represented around 60 per cent of the total volume of crude processed). This refining pattern was the result of two conditions. First, the 1943 Hydrocarbons Law (drafted with the purpose of forcing the concessionaires operating in Venezuela to refine more of their crude within the country) made the installation of very large and simple refineries a profitable undertaking for the foreign oil companies, mainly because tax concessions were offered to companies based on the amount of oil they refined, rather than on the sophistication of the refining process they used (Philip, 1982, p. 423). Secondly, the growing demand for fuel oil which for many years characterized the products market of the eastern seaboard of the USA meant that heavy fuel oil could be marketed in large quantities by the concessionaires on the US East Coast. By 1976, however, this refining pattern had become a liability for PdVSA. The increasing domestic demand for light products – especially gasoline – forced the company to produce very large volumes of fuel oil at any cost, just to be able to satisfy this internal demand. This reduced the operating flexibility of the company to a minimum: on the one hand, PdVSA had to place large volumes of fuel oil overseas, precisely at a time when the market for this product was far from buoyant; on the other hand, it had to increase runs of valuable (and scarce) light and medium crudes, so as to cope with periods of low fuel oil demand while still turning out the necessary light ends.[2] Finally, the company found itself facing an octane deficit, due to insufficient installed capacity in terms of catalytic reforming and catalytic cracking (in 1976, only the Cardón and Puerto La Cruz refineries had any conversion capability). Lead usage, as a consequence of this, had risen to the maximum tolerable level of 3 m^3/gallon.

Oil industry planners in Venezuela had identified these problems a long time before the nationalization took place, but they had not had the power to modify the refineries. Naturally, when the nationalization occurred, PdVSA began the implementation of a comprehensive rationalization programme, aimed at ridding the refining system of its flab by shedding PdVSA's obsolete refining assets, as well as a massive investment programme, whose objective was the modernization of the country's refining plant and equipment.

The rationalization effort got under way almost immediately after the nationalization, with the administrative and logistical integration of the Puerto La Cruz and El Chaure refineries. In 1977, the shutting down of the Mobil refinery at Caripito marked the start of a major streamlining process, which would see the Tucupita, Morón, San Lorenzo and Bajo Grande refineries ceasing their operations, in the years 1978–9, 1981, 1982 and 1987, respectively.[3]

The investment programme consisted of a number of projects which sought to increase gasoline yields at the expense of fuel oil yields; to maintain processing levels intact, while replacing some of the light and medium feeder crudes with heavier crudes (so as to produce greater amounts of high value light ends from abundant heavy crudes, instead of the then-pervasive situation of transforming valuable light and medium crudes into low value residual fuel oil[4]); to diminish the use of lead as an octane enhancer, and finally, to improve the quality of residual fuel oil. The great importance which PdVSA accorded to its refinery modernization programme is better understood by looking at Figure 6.1, which shows how its investments in refining built up much more rapidly than those in exploration, production or marketing.

The first refinery upgrading programme to be implemented was called ELPAEX, and its subject was the El Palito refinery. From 1976 onwards, the refinery received a 42,000 b/d fluid catalytic cracking unit and 22,000 b/d HF alkylation unit. This enabled it to raise its gasoline output from 17,000 b/d to 77,000 b/d,[5] to produce enough octane to cut lead usage to 1 m^3/gal, and finally, to reduce its residual fuel oil output from 60,000 b/d to 29,000 b/d; all of this by 1982. After El Palito came Cardón, where Maraven revamped the FCC unit, and added an isomerization and alkylation complex, a lube plant, and a pilot hydrodemetallization plant. But by far the single biggest item in PdVSA's refining programme was the upgrading of Lagoven's Amuay refinery, begun in 1977 and known in Venezuela as the MPRA project (*Modificación del Patrón de Refinación de la Refinería de Amuay* or Modification of the Refining Pattern of Amuay Refinery). This project saw the installation

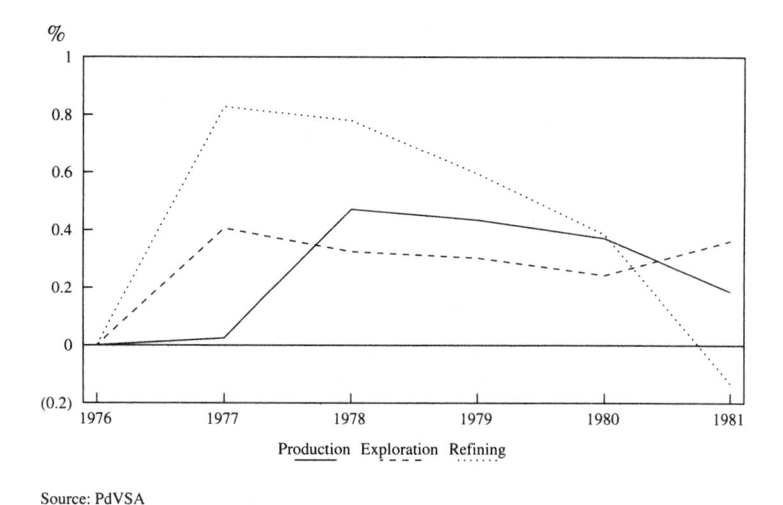

Figure 6.1: PdVSA Investments by Activity. 1976–1981.

of a 52,000 b/d flexicoker (by far the biggest in the world at the time), a 74,000 b/d flexicracker (the biggest cracking unit in Latin America) and associated isomerization and alkylation plants. The MPRA project, costing about $770 million, was extremely innovative at the time, in the sense that its centrepiece was a type of processing plant (the flexicoker) which had never been built on such a large scale before.[6]

The El Palito and Amuay projects came on stream just in time to save PdVSA from what, due to the crumbling international demand for residual fuel oil since 1981, could have been an extremely tight (and embarrassing) domestic supply situation. The company got a very good picture of 'what could have been' had it not embarked upon the upgrading programme when, just as El Palito was due to come back on stream, the international fuel oil market collapsed. This forced PdVSA to reduce its refining operations to a bare minimum, while importing gasoline from Trinidad in order to guarantee local supply (*OB*, July 1982, p. 74).

Unfortunately, the soft oil market of the early 1980s also prompted PdVSA to shelve some of its other refinery upgrading plans. The most important of these involved the Puerto La Cruz refinery, which was to have received a new 52,500 b/d fluid catalytic cracker, and two new vacuum units, as well as being reconfigured to be able to process around 25,000 b/d of heavy crude, all at a cost of some $730 million (*O&GJ*, 25 August, 1980, p. 72). Another abandoned scheme called for a 100,000 b/d facility with atmospheric distillation, high vacuum distillation, deasphalting, hydrotreatment and hydrocracking plants, which was to have been installed at Cardón. This complex would have enabled Cardón to deasphalt Tía Juana Heavy residual, hydrotreat it using a catalyst with a high metals uptake capacity, and finally, hydrocrack it in two trains of about 25,000 b/d capacity, in order to produce diesel. Yet another important project which did not see the light of day was a pilot 2,500 b/d demonstration unit of UOP's Aurabon process, which was to have been installed at the Bajo Grande refinery. This process, according to its inventors, enables a refiner to convert, via catalysis, a heavy feedstock into a product with much reduced asphaltene and metal contents. The plant's charge would have been vacuum residual from 10°API Boscán crude (produced in the vicinity of this refinery). Preliminary tests had shown very encouraging results: the metals content was reduced from 1,484 ppm in

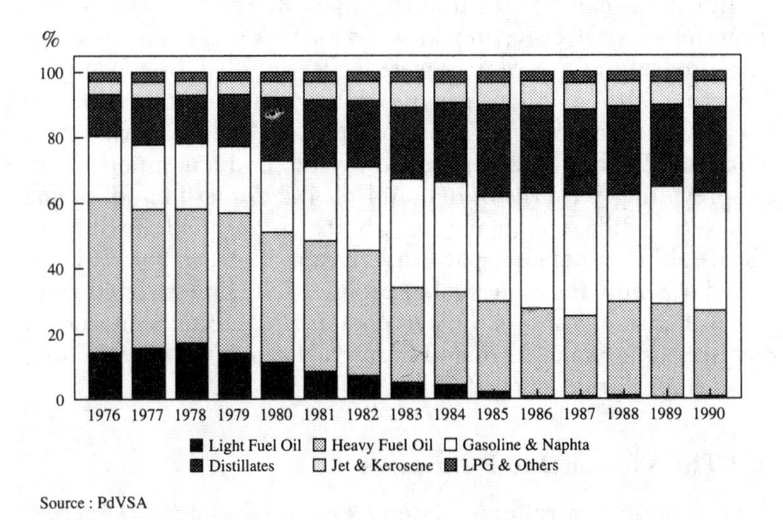

Source : PdVSA

Figure 6.2: Product Slate of the Venezuelan Refineries.

Figure 6.3: Utilization of Refinery Capacity in Venezuela. 1976–90.

topped Boscán to 54 ppm in the 566°C+ portion of the product after treatment, with the product yield at 343°C+ amounting to 89.2 per cent of total volume.[7]

In sum, it can be said that the 1976–82 modernization programme was a decisive step forward for PdVSA. Even if some of the programme's promises were never fulfilled (for example, lowering the average gravity of Amuay's crude diet to 21°API), there can be little doubt that 'the modernization enabled PdVSA to bring its supply of products much closer to prevailing patterns of demand on the domestic and export markets, while enhancing its ability to respond flexibly to future shifts in demand patterns' (Evans, 1991, p. 388). Figure 6.2 shows how the Venezuelan product slate has changed since 1976; Figure 6.3 shows the percentage of domestic Venezuelan refinery usage since 1976 and Figure 6.4 shows the way refining costs have varied since 1975.

6.2 The Venezuelan Refineries

The Venezuelan refining system is composed of six refineries: Amuay, Cardón, El Palito, Puerto La Cruz, El Toreño and

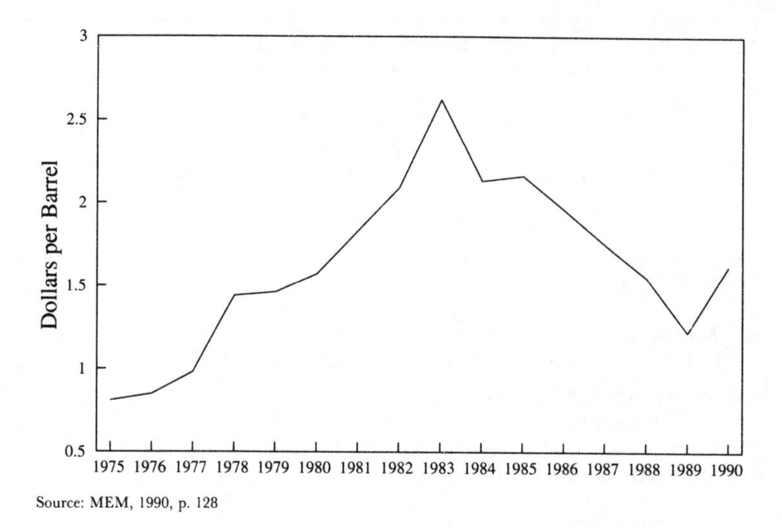

Figure 6.4: Refining Costs in Venezuela. 1975–90.

San Roque; their major characteristics are summarized in Tables 6.2a, 6.2b, 6.2c and 6.2d.

The *Amuay* refinery[8] which commenced its operations in 1950 was built by the Creole Petroleum Corporation (a subsidiary of Jersey Standard Oil). Originally, it had a 60,000 b/d capacity,

Table 6.2a: The Venezuelan Refining System. Charge Capacity in Barrels per Day.

	Atmospheric distillation	*Vacuum distillation*	*Catalytic cracking*	*Catalytic reforming*	*Merox*	*Alky-lation**
Amuay	635,000	373,500	80,000	-	83,800	17,000
Cardón	300,000	156,500	63,000	-	52,300	25,000
Puerto La Cruz	199,000	-	12,000	-	6,200	2,200
El Palito	110,000	66,500	48,000	6,500**	45,000	22,000
El Toreño	5,000	-	-	-	-	-
San Roque	5,300	1,800	-	-	-	-
TOTAL	1,254,300	598,300	203,000	6,500	187,300	66,200

* Capacity expressed in terms of production
** When operating at high severity (98 RON). At low severity (94 RON), capacity is 7,500 b/d.
Source: CEPET, 1989, v. I, p. 558

Table 6.2b: The Venezuelan Refining System. Capacity in Barrels per Day.

	Viscosity reduction	*Desulphurization (Light distillates)*	*Desulphurization (Heavy distillates)*	*Isomerization**	*Coking*	*HDH***
Amuay	-	69,000	170,000	8,900	55,000	-
Cardón	90,000	36,500***	35,000	11,000	-	2,500
Puerto La Cruz	-	-	-	-	-	-
El Palito	-	-	-	-	-	-
El Toreño	-	-	-	-	-	-
San Roque	-	-	-	-	-	-
TOTAL	90,000	105,500	205,000	19,900	55,000	2,500

* Capacity expressed in terms of production
** Hydrodemetallization
*** Capacity processing kerosene. When processing gasoil, capacity is 36,000 (32,500 b/cd), and when processing light cycle oil, capacity is 23,500 (21,120 b/cd).
Source: CEPET, 1989, v. I, pp. 558–9

Table 6.2c: The Venezuelan Refining System. Capacity in Barrels per Day.

	*Lubes**	*Asphalt*	*Paraffin+*	*Sulphur+*	*Hydrogen***
Amuay	2,300	53,100	-	600	93
Cardón	5,800	-	-	94	27
Puerto La Cruz	-	-	-	-	-
El Palito	-	-	-	36	-
El Toreño	-	-	-	-	-
San Roque	-	-	79	-	-
TOTAL	8,100	53,100	79	730	120

* Capacity expressed in terms of production
\+ Tonnes per day
** Millions of cubic feet per day
Source: CEPET, 1989, v. I, p. 559

but subsequent modifications transformed it into the largest refinery in the world, with a nameplate capacity of 670,000 b/d. Until 1976, Amuay's sole purpose was to produce what can only be described as colossal amounts of fuel oil. Located to the west of the Paraguaná peninsula, in the state of Falcón, Amuay is presently operated by Lagoven. The location has one important drawback; namely, that water has to be brought

Table 6.2d: The Venezuelan Refining System. Industrial Services and Storage Capacity.

	Electricity	*Steam*	*Crude storage*	*Products storage*	*Fuel oil storage*
	(MW)	*(T/hr)*	*000 barrels*	*000 barrels*	*000 barrels*
Amuay	140	1,680	4,299	11,695	29,225
Cardón	89	480	3,381	7,244	11,496
Puerto La Cruz	37	215	6,240	3,974	2,739
El Palito*	39	436	1,200	2,760	1,528
Bajo Grande**	0	59	5,430	1,342	2,342
San Roque	2	32	55	47	10
TOTAL	307	2,902	20,535	27,062	47,937

* Includes El Toreño
** Not operated as a refinery
Source: CEPET, 1989, v. I, p. 566

by pipeline from the reservoirs of El Isiro and Barrancas, 140 km. away. During the summer, when water levels in these reservoirs are low, water has to be transported by tanker all the way from the Caroní River (Bolívar state), in order to satisfy the huge water needs of the refinery (CEPET, 1989, v. 1, p. 565). Amuay has a natural harbour which faces the Gulf of Venezuela. The refinery receives crude by means of two pipelines with a 760,000 b/d capacity, which originate 235 km. away, in Ulé, state of Zulia. Amuay can also receive crude by means of tanker ships. At the end of 1991, PdVSA completed an integration project linking the Amuay and Cardón refineries, which will help to optimize operations at both plants (*WPA*, 27 January, 1992, p. 19). The volumes and quality of the crudes processed in the refinery, naturally, determine the pattern of product yields. Low throughputs (320,000 b/d) permit maximum conversion capacity utilization, with higher throughputs resulting in increases in the yields of high and low sulphur fuel oil. The present pattern of yields at Amuay, under different crude diets and capacity utilization conditions, is illustrated in Table 6.3. The refinery supplies the Guayana and Caracas zones, as well as the western parts of Venezuela, with petroleum products. Furthermore, it has a sizeable exportable surplus.

The *Cardón* refinery was built by the Royal Dutch/Shell

Table 6.3: Amuay Refinery: Product Yield With Different Operating Diets.

Capacity utilized	*320,000 b/d*	*500,000 b/d*
Feeder Crudes (% volume)		
Light	22	5
Medium	88	75
Heavy	-	20
Products (% volume)		
LPG	2	1
Naphthas and gasolines	32	25
Distillates	34	28
Low sulphur fuel oil	10	14
High sulphur fuel oil	8	22
Asphalt	12	8
Lubes	1	1
Own consumption	1	1

Source: CEPET, 1989, v. I, p. 564.

company. Its initial capacity was 30,000 b/d but, as a result of various enlargement programmes, the present nominal capacity of the refinery stands at 369,000 b/d. Cardón is located in the same peninsula as Amuay, and its water supply logistics and problems are similar. Cardón has long been considered the most flexible refinery in Venezuela, as it has always been able to turn out a very wide variety of petroleum products (around 250). It supplies the Caracas and western Venezuela markets, and its products surpluses are exported to the United States, Europe and Japan. Cardón receives its light crude feedstocks from the Maracaibo region via a pipeline with a capacity of 330,000 b/d. Natural gas is sent to Cardón from Puerto Miranda in a 62 mcf/d pipeline. A propane pipeline brings this gas from Amuay to Cardón. Heavier crudes, as well as vacuum residue, vacuum distillates and butanes, are brought to the refinery by tanker. The present pattern of yields at the refinery is illustrated in Table 6.4.[9]

El Palito refinery was built by Mobil in Carabobo state, within easy reach of the main industrial centres of the country. Originally, its capacity was 55,000 b/d, but it now stands at 110,000 b/d. Its strategic location makes it the choice provider of fuels for the central states of Venezuela. The crudes processed at El Palito come from Barinas, via a 540 kilometre-long

Table 6.4: Cardón Refinery. Product Yield With Different Operating Diets.

Capacity utilization	*234,000 b/d*	*215,000 b/d*	*315,000 b/d*
Feedstocks (% volume)			
Light/medium crude	64	97	92
Heavy crude	21	-	5
Butane	2	2	2
Propane	1	-	1
Natural gasoline	1	1	-
Long residue	11	-	-
Products (% volume)			
LPG	1	1	1
Naphthas and gasolines	33	38	31
Distillates	26	34	35
Heavy fuel oil	34	21	25
Lubes	2	2	1
Own consumption	4	4	7

Source: CEPET, 1989, v. I, p. 577.

pipeline (longest in the country), which has a 100,000 b/d capacity. It also receives crudes and atmospheric residue from the Punta de Palma terminal by means of tanker ships. Isobutane for its alkylation plant comes from the LPG plants at Ulé and Bajo Grande (Zulia state). El Palito is very important from the point of view of product distribution in the Venezuelan market. The refinery has a number of filling stations for road tankers, and two products pipelines connect it with the main products distribution centres which PdVSA has in the region. The operations of another Corpoven refinery, *El Toreño*, are partially integrated with those of El Palito. The purpose of this arrangement is to maximize the use of the conversion capacity installed in the bigger refinery. El Toreño is located in the southwesterly part of Barinas state. The atmospheric residue produced in this refinery is sent, together with some Barinas crude, by pipeline to El Palito, where it is processed. Table 6.5 indicates the typical operational balance of the El Palito/El Toreño refineries, before the expansion of the fluid catalytic cracker capacity at El Palito to 54,000 b/d.

The *Puerto La Cruz* refinery was built by Venezuelan Gulf Oil. Completed in 1950, its original capacity was 30,000 b/d. Its distillation capacity has increased significantly since then; its

Table 6.5: El Palito and El Toreño Refineries. Product Yield With Different Operational Diets.

Feedstocks (% volume)		
Light crude	75	46
Heavy crude	-	30
Others*	25	24
Gravity (°API)	29	23
Products (% volume)		
LPG	2	2
Naphthas and gasolines	50	39
Distillates	19	19
Heavy fuel oil	27	37
Own consumption	2	3

* Long residue, isobutane and straight run gasoline
Source: CEPET, 1989, v. I, p. 584.

present nameplate capacity is 195,000 b/d. The refinery at El Chaure (built by Sinclair in 1950 in the vicinity of the Gulf refinery) is administratively and logistically integrated to Puerto La Cruz. Pipelines which permit the transfer of refined fractions between the refineries make it possible to operate them as a single unit. Puerto La Cruz supplies the Venezuelan internal market with LPG, gasolines, kerosene, diesel and fuel oil. Products surpluses from the refinery are either exported or sent to other refineries for reprocessing. The crudes processed in Puerto La Cruz come from the state of Anzoátegui through five pipelines. Isobutane for its alkylation unit originates at the natural gas liquids fractioning plant at Jose. Filling stations for road tankers are found in Guaraguao and El Chaure. The products yield of this refinery, under different crude diets, is shown in Table 6.6.

The *San Roque* refinery was built by Phillips Petroleum in 1952, in order to produce paraffins. Its original capacity of 2,100 b/d has multiplied roughly two-fold, and now stands at 4,500 b/d. It is still the only refinery producing paraffins in Venezuela, and its output is utilized almost exclusively by candle manufacturers. San Roque runs a mixture of crudes with high paraffin content, which come from fields in the central zone of Anzoátegui state (specifically, the Santa Ana, San Joaquín and El Toco fields).

Table 6.6: Puerto La Cruz Refinery. Product Yield With Different Operational Diets.

Feedstocks (% volume)		
Light crude	72	49
Heavy crude	28	51
Gravity (°API)	30	24.8
Products (% volume)		
Naphthas and gasolines	30	22
Distillates	21	17
Light fuel oil	9	3
Heavy fuel oil	39	57
Own consumption	1	1

Source: CEPET, 1989, v. I, p. 589.

6.3 Outlook: Greener Products, Dirtier Crudes

Without doubt, one of the biggest problems facing the international refining industry nowadays is turning out ever greater amounts of so-called 'green' products (reformulated high octane gasoline with a low Reid Vapour Pressure and lower carbon dioxide emissions, or low sulphur diesel, for example). Coping with the increasing demand for environmentally friendly products is no easy task, as Seymour (1992, *passim.*) has shown in his study of the impact of reformulation regulation on the US refining industry. It requires sizeable investments to upgrade existing refinery capacity in order to meet progressively tougher environmental standards and products specifications.[10]

The 'greening of the barrel' is a process which PdVSA, one of the world's biggest refiners, also has to face (especially since it exports large quantities of products to the United States, a country with some of the most stringent environmental regulations in place). But the company's situation is even less enviable than that of other refiners, because its response to the green challenge will have to come at a time when the diet of its refineries is undergoing a progressive shift towards heavier, sulphurous, crudes.[11] PdVSA knows that in order to bring down fuel oil yields to, say, about 12 per cent of total product yield, while running twice as much crude with a gravity inferior to 22° API, its domestic refining system will require massive modifications. Thus, the company's budget plan for the 1991–6 period set aside $8 billion to cover its investments in refining.

The largest project contemplated in this budget, the construction of a 230,000 b/d grass-roots refinery near Jose, was to be carried out by its future operator, Corpoven. The plant, known acronymically as NRO (*Nueva Refinería de Oriente*, or New Eastern Refinery), would have cost anything from $3 to 5 billion, and it was scheduled to come on stream in 1997; follow-on plans call for its expansion to a capacity of 400,000 b/d by the year 2000 (*WPA*, 27 January, 1992, p. 19). The refinery was to have two distillation and vacuum units, and two flexicokers of 55,000 b/d capacity each. Output from the first stage of the NRO project was planned to consist of a majority of light ends (85 per cent), with only a 15 per cent yield of residuals.[12] Feedstock for the new plant was to have been 10° API crude from the Hamaca area of the Orinoco Oil Belt.[13] This crude would have had to be mixed with kerosene at the production site (this will give the oil a 15° API gravity), in order to be moved by pipeline to the refinery. To supply the refinery, Corpoven was to build two new pipelines with a total length of about 170 km. each; the first would have transported the diluted crude from Hamaca to the refinery, while the second would have been used to recycle the kerosene to the production sites.

Originally, Corpoven planned to cover the construction costs of the refinery alone (using a combination of its own capital, and project financing credits obtained from international financial institutions), but the price tag attached to the NRO projects, and the drastic reduction which the government effected on PdVSA's budget in 1992, has forced PdVSA to find foreign partners which, in return for equity participation in refining ventures, are willing to shoulder the financial burden of building it.[14] No one has yet stepped forward to do this, because potential investors have made it clear that they would consider participating in such a project in the future if, and only if, the income tax rates for oil activities in Venezuela were revised downwards, and the government established guarantees that would prevent it from altering these rates at whim. In any case, the likelihood of the NRO project going ahead should be considered very doubtful.

In 1991, Venezuela's effort to attract foreign investment to the refining sector received an important boost when the minister of energy and mines at the time, Celestino Armas, let it be known that the Pérez administration would enact a special reduction that would leave taxes on refining activities at around 35 per cent, a level similar to the tax levied on PdVSA's petrochemical operations and joint ventures (*O&GJ*, 26 August, 1991, p. 35). Critics of this measure pointed out that it did not go far enough in remedying the tax structure which is strangling PdVSA, and that the new law would only discourage the company from making refining investments on its own, since joint ventures with private investors are eligible for a lower tax ceiling (*PON*, 20 November, 1991, p. 5). This, plus the recent cuts in PdVSA's budget, have prompted rumours suggesting that the company is considering opening projects that will add incremental refining capacity in its existing refineries (like the ones detailed below) to foreign investment.

Maraven plans to expand its refining capabilities through four different projects, all located at its Cardón refinery. The objective is to enhance the refinery's capacity to produce a slate of higher quality products for export purposes. The financial outlays associated with these projects are estimated at $1.8 billion over the years 1991–6. The first project involves the installation of a 60,000 b/d delayed coker, scheduled to go onstream at the end of 1994. This unit is intended to reduce Cardón's residual fuel oil output (currently hovering around 25 per cent) to a minimum level, only sufficient to cover domestic needs for this product. The fuel grade coke produced by the plant is to be sold to cement producers in Europe (*O&GJ*, 26 August, 1991, p. 36). The second project calls for a 45,000 b/d naphtha reformer, to come on stream at the end of 1994. This plant would enable Maraven and Pequiven to build, in due course, an aromatics plant at the Cardón compound. The objective of the third project, a 15,000 b/d isomerization unit, is to boost the octane levels of the refinery's gasoline pool. Potentially, however, the most important (and ambitious) of Maraven's plans for Cardón is a projected 15,000 b/d hydroprocessing unit. This plant will use the hydrogenation and hydroconversion (HDH) process developed by INTEVEP (which won PdVSA's research arm an UNESCO prize). Cardón's HDH plant would be one hundred times bigger than the largest HDH plant currently under operation, a 150 b/d unit located at Ruhr Öl's Scholven refinery, and PdVSA hopes that it will also serve as a technology demonstrator, convincing prospective buyers abroad of the great potential of the process. PdVSA claims that HDH can reduce the sulphur and aromatics content in diesel, and that it can upgrade cracker feed from deep conversion units (thus making it suitable for processing Orinoco crudes).[15] Cardón is also to receive MTBE and TAME units.

The Amuay refinery, the jewel in Lagoven's crown, will receive a 28,000 b/d delayed coker in 1994, as well as a fractioning unit which will allow increased production of low sulphur fuel oil from late 1993 onwards. To this will be added MTBE and TAME units, a diesel fuel recovery system, a coke flue gas treating unit and a sulphur recovery unit. Also on the cards is an interesting plan which calls for a change in the refinery's crude diet. According to this plan, the quality of the average crude run in Amuay is to fall from 25.8° API to 23.5° API by 1996. This would involve increasing the light crude run from 450,000 b/d to 500,000 b/d, and doubling the heavy crude run from 50,000 b/d to 100,000 b/d.[16] Such an action would add around 40 per cent to refining costs. The start-up date for the shift, however, has had to be delayed until 1993 (instead of beginning in late 1991); according to the *Energy Economist*, the plant manager attributed the delay to the fact that the refinery had to 'meet the market' (*EE*, December, 1991, p. 18), and could not afford to make such a move at that time. This illustrates vividly the difficulties inherent in refining some of the toughest crudes in the world in order to produce green fuels. However, after considering PdVSA's refining plans for the remainder of the century, it is clear that the Venezuelans appreciate the necessity to invest heavily to cope with the environmental quality requirements which many governments are now imposing on petroleum products.

Currently, the other main area of concern for PdVSA's refining plans is one where upstream and downstream can be said to overlap: the exploitation of the Orinoco Oil Belt. As has been shown, PdVSA expects that a number of foreign oil companies will build deep conversion refineries to upgrade Orinoco crudes, in exchange perhaps for equity holdings in more attractive areas of the Venezuelan upstream. The prospects for this scheme are discussed elsewhere in this study.[17]

Notes

1. As the *Oil & Gas Journal* put it: 'PdVSA inherited the wrong tool to meet a demand pattern [which was] almost the opposite of the refinery output pattern'(*O&GJ*, 25 August, 1980, p. 70).
2. Even so, the production of fuel oil in the country was so large that many times PdVSA had to resort to 'filling huge earthen reservoirs with resid in the summer – in the hope that winter demand in North America or elsewhere, would empty them' (*O&GJ*, 25 August, 1980, p. 71).
3. PdVSA's original streamlining plans contemplated shutting down the San Roque and El Toreño refineries by 1983. The Bajo Grande refinery was not originally meant to cease operations (See *PON*, 23 September, 1982).
4. In 1976, 90 per cent of Venezuela's light crude production was run through the country's domestic refining system.
5. Not all of these gains were due to the new processing units, however. 17,000 b/d of natural gasoline from El Toreño refinery also helped to boost the yield.
6. To give an idea of what the new unit was supposed to contend with, it is sufficient to say that its feed was to contain an unprecedented 708 ppm of vanadium and 163 ppm of nickel. This gave the coke from the flexicoker (produced at a rate of 100 tonnes per day) a staggering 7 per cent vanadium content by weight. If it had all been separated, the vanadium from all this coke would have been equivalent to 10 per cent of the world demand for this metal (*O&GJ*, 15 September, 1980, p. 181). Nowadays, production of coke at Amuay is up to 360 tonnes per day (*O&GJ*, 13 April, 1992, p.40).
7. This project offered a very good use for Boscán residual which, unlike, say, Orinoco crudes, could not find an economic outlet as a fuel for steam generation, since the Boscán formations are too deep for steam stimulation.
8. Sometimes referred to as the Judibana refinery.
9. An experimental residue hydrodemetallization complex operates – on a semi-commercial basis – at Cardón. The demetallized residue obtained in this complex is mixed in the fuel oil pool of the refinery.
10. Reformulating automotive fuels, for instance, will have a cost for US refiners estimated at anything between $25 – 33 billion (Seymour, 1992, p. 60).
11. Production plans for 1996 call for boosting the amount of under-22° API crudes to be processed by 32 per cent (up to more than 1 mb/d). By 1996, PdVSA expects the flow of heavy crude in its domestic and Curaçao refineries to be 650,000 b/d (*PIW*, 4 November, 1991, p. 8).

12. The refinery will also generate about 1,000 tonnes per day of sulphur, which Corpoven plans to export. Following the arrangement pioneered at Amuay, coke from the flexicokers will be burned to generate steam, and flexigas will be used as a refinery fuel.
13. In addition to the 230,000 b/d of crude capacity, however, the plant will also be able to run 47,000 b/d of residual from the Puerto La Cruz refinery.
14. In fact, as *PON* reports, the NRO project 'will only be carried out if foreign oil companies assume *majority* equity and control of the project, leaving PdVSA with only small participation' (26 May, 1992, p. 5).
15. According to *PIW* (25 November, 1991, p. 3), this pilot project had been postponed for budgetary reasons months before the large cuts in PdVSA's budget were announced.
16. For the country as a whole, the aim is to run 500,000 b/d of heavy and extra-heavy crude by 1997.
17. At the time of writing (1992), among the negotiations which PdVSA is holding with different foreign oil companies, those with the German company Veba were the most advanced. In May 1992, executives from the two companies met in order to review a pre-feasibility study for a 100,000 b/d conversion refinery. In the fall of 1992, they will decide whether to go ahead with engineering studies for the project, whose estimated cost is $4 billion. According to Veba, quick approval of the project could make possible a 1996 start-up date, but this is highly unlikely.

7 THE MARKETS FOR VENEZUELAN PETROLEUM

7.1 Crude Oil Exports

Venezuela began exporting crude almost as soon as the first commercial oil deposits were found in Lake Maracaibo. Since then, about 90 per cent of all the crude ever produced in the country has been exported (as crude, or in the form of products), because the satisfaction of Venezuela's internal energy needs has never required the use of more than a small fraction of the country's total output.

Until 1976, Venezuela's crude exports were mainly light varieties with a gravity greater than 30° API. However, after 1977, light crude exports began to decline rapidly, reaching their nadir in 1984, when their volume accounted for only 12 per cent of Venezuela's export flows. Medium crudes exhibited a similar tendency: in 1985, their share fell to an all time low of 10 per cent. The discovery of new light and medium crude fields after 1985 reversed this trend, so that, once again, high value light and medium crudes have come to account for a majority of Venezuela's crude exports.

Historically speaking, the majority of Venezuelan petroleum exports has gone to the US market, either directly or indirectly (in the form of products processed in Aruba and Curaçao from Venezuelan crude). This dependence has made Venezuela particularly vulnerable to any action on the part of the USA hampering the flow of Venezuelan oil into the American market. Unfortunately for Venezuela, the American congress, under pressure from the very powerful lobby of American oil producers, has at various times enacted legislation restricting the free trade of oil to the USA. In 1932, for instance, it imposed a tariff on imported oil; as we have seen, this action led to Standard of Indiana's sale of its Venezuelan assets to Standard of New Jersey. This threat to Venezuela's livelihood was ended in 1935, when the 1932 law was declared unconstitutional. However, the rising level of imports during the early 1950s prompted producers in Texas, Louisiana and Oklahoma

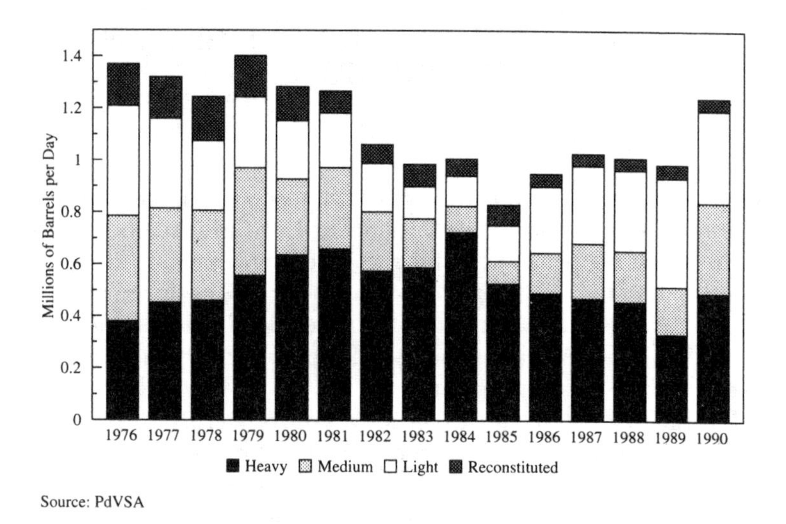

Source: PdVSA

Figure 7.1: Venezuelan Exports of Crude Oil by Type, 1976–90.

to protest anew, demanding protection for the domestic oil industry on grounds of 'national security'. This prompted Dwight Eisenhower to appoint a commission to study the issue. The commission decided that import quotas were unnecessary, and that national security interest would be best served if oil importers exerted voluntary restraints in order to maintain the ratio of oil imports to production at the level it had in 1954, i.e. about 10 per cent of total consumption (Danielsen, 1982, p. 147). In the years 1957–9, however, imports rose sharply once again, and shut-in production capacity reached a level of 38.1 per cent. In response to these new developments, Eisenhower issued an executive order authorizing the imposition of mandatory import controls. The main consequences of the Mandatory Oil Import Control Program (MOIP) was that it cut off the USA from the rest of the world oil market,[1] and this in turn led to 'reduced demand for oil from foreign countries . . . downward pressure on world oil prices[2] and increased capacity utilization rates in the United States' (ibid., p. 150).

The effect which MOIP had on Venezuela's oil exports was far reaching: the country's share in American crude imports fell from 47 per cent of the total volume in 1959 to only 20 per cent in 1970 (from 450,000 b/d to 268,000 b/d). Not surprisingly, the Venezuelan government tried by all means to get preferential treatment from the US government, but its efforts proved fruitless. The import quota episode left a legacy of mistrust against the USA in the mind of many Venezuelan politicians, and it was also one of the factors that led Venezuela to propose the foundation of OPEC.

As Table 7.1 shows, the USA is still Venezuela's most important customer. Flows of Venezuelan crude to Europe, however, have grown noticeably since the beginning of the 1980s, as a result of PdVSA's increasing downstream presence in the zone. In the foreseeable future, the regional breakdown of Venezuela's exports is not expected to change drastically. In 1991, for instance, PdVSA announced that, by the year 2000, its crude exports to Latin America and the Caribbean would reach 600,000 b/d, while her exports to Europe would be around 400,000 b/d. However, the USA, accounting for around 2 mb/d of Venezuelan crude, would still be the mainstay of Venezuelan oil trade (*PON*, 14 November, 1991, p. 3). These predictions, however, were made on the basis of total Venezuelan production reaching a level of 5 mb/d by 2000. Since the recent cuts in PdVSA's investment plans will make the attainment of these output goals extremely unlikely, these estimates will have to be revised downwards by a considerable amount.

There is an aspect of Venezuela's crude oil trade which deserves a special mention (since it plays a crucial role in the

Table 7.1: Venezuela. Principal Destinations of Crude Exports, 1986–90. Barrels per Day.

Destination	*1986*	*1987*	*1988*	*1989*	*1990*
Total Exports	948,700	1,026,800	972,700	986,300	1,241,900
Canada	35,200	34,900	30,500	40,100	30,800
USA	425,500	479,600	402,900	586,500	653,900
Latin America	295,100	339,900	328,300	218,600	335,600
Germany	93,000	83,300	83,600	76,100	78,200
Other Western Europe	93,600	82,400	77,100	57,700	130,400
Japan	4,400	4,800	5,200	6,000	6,000

Source: OPEC, 1990, p. 87.

promotion of the country's foreign policy objectives in Latin America): her supply agreements with Central American states. The first of these, called the Puerto Ordaz agreement, was signed by Venezuela and six states – Costa Rica, El Salvador, Guatemala, Honduras, Nicaragua and Panama – in December 1974. These nations received oil from Venezuela under quite favourable financial terms: they were charged $6 for every barrel of crude they received; the monetary difference between this number and the prevailing official price at the time of the purchase was treated as a long-term government-to-government loan (Evans, 1991, p. 385). The Puerto Ordaz agreements were superseded by the San José accord (August 1980), concluded jointly between Venezuela and Mexico for the supply of 160,000 b/d of crude to the six countries mentioned above, plus Barbados, Jamaica, the Dominican Republic, Belize and Haiti. In its original form, the San José accord stipulated that 30 per cent of the receipts from the sale of this oil at official prices was returnable to the importing governments in the form of loans repayable over 5 years at 4 per cent interest. These very favourable terms, however, did not survive the 1982 oil price fall (which brought about a near collapse in the economies of both Mexico and Venezuela). The 'returnable' portion of the oil sales covered by the agreement was cut from 30 per cent to 20 per cent, and the credit terms were also altered (new loans were to be repaid over five years at 8 per cent interest, with the option of conversion to 20 years' repayment at 6 per cent interest if the funds were invested in development projects).[3]

Under the terms of the San José accord, Venezuela and Mexico have supplied, on a near-equal basis, a total of over 300 mb of oil to their neighbours in Central America and the Caribbean since 1980, assuring around half of the 225,000 b/d petroleum requirements of the beneficiary countries (*PIW*, 30 July, 1990, p. 5). In return for supplying this cheap oil, Mexico and Venezuela expected to see their political leverage within the region grow. And indeed, in the early years of its existence, the San José accord proved to be a very potent and high profile foreign policy instrument.[4] As a result of the fall in oil prices of the early 1980s, however, obtaining hard cash has become a more pressing priority than scoring foreign policy points. Thus, both countries have adopted a more pragmatic attitude towards the question of using the accord to bolster their image.[5] Nevertheless, many critics of the Venezuelan government's policies have demanded that the accord be stopped, calling it a waste of government resources (*PON*, 1 June, 1992, p. 3).

7.2 Term Contracts for Crude Oil

Venezuelan crude term contracts can be said to be options – rather than obligations – to buy a specified volume of a specific crude grade every *x* period of time. This makes them very different from the 'typical' term supply contracts used by other producing nations. We can illustrate this point by using the example of Mexico. In a Mexican crude contract, the customer has the obligation to lift the specified volume of crude every time *x* days have passed, and PEMEX, in turn, has to deliver the specified volume and grade of crude.[6] A Venezuelan contract, however, gives the buyer discretion on whether or not to lift the oil in any given month, without the need to declare *force majeure*, or to forego the possibility of buying oil from PdVSA in the future, should he choose not to lift. In other words, a Venezuelan contract gives the customer the right to lift oil should he find the price attractive at any particular moment in time; the customer is only bound to lift a specified (variable) percentage of the total volume covered by the contract during the year. Of course, this bonus has a catch attached to it: just as a customer can, if he so chooses, buy less crude than the volume specified on the contract, PdVSA's operating subsidiaries can, at their discretion, supply volumes inferior to those specified.

Up until 1976, the commercialization of Venezuela's petroleum exports was basically a province reserved to the concessionaires operating in the country.[7] As a consequence of this, after the oil nationalization, PdVSA found itself in a rather uncomfortable situation, as far as marketing its crude output was concerned, because its dependence on a relatively small base of contractual clients (mainly its ex-concessionaires) was excessive. In the negotiations that followed the expropriation of their assets, the ex-concessionaires drove a very hard bargain with the newborn PdVSA, threatening to refuse to lift Venezuelan oil unless they virtually dictated its price. The Venezuelans were powerless to stop the companies, because their lack of commercial expertise would have seriously handicapped any short-term effort on their part to place their crude in alternative markets.[8] Quite logically, PdVSA sought to emancipate itself from the thrall in which these companies held it by finding new customers for its oil. This endeavour was made substantially easier by the tightening market situation which prevailed at the end of the 1970s, and which came to a head in the aftermath of the 1980 Islamic revolution in Iran. The sellers' market prevailing in 1979 greatly strengthened Venezuela's hand when the time to renegotiate PdVSA's sales contracts came around (1979). In this round of negotiations, PdVSA agreed to continue selling oil to its ex-concessionaires, but on the understanding that they could change the contractual terms at only one month's notice. Furthermore, it imposed a standard contract on all its customers (in other words, it stopped granting special allowances to some). The duration of the new contracts was set at one year (except in the case of contracts for heavy and extra-heavy crudes, which are valid for two years). The contracts continued to be renewable by mutual consent of the parties, but they contained a 'destiny clause' giving PdVSA the right to decide on who the final customers for its oil could be.

PdVSA's efforts to diversify its contractual clients base has been quite successful, as Table 7.2 shows. In 1976, the liftings of Venezuela's ex-concessionaires accounted for about 80 per cent of all of the country's exports; by 1985 their share was down to 32 per cent. After this year, and as a result of PdVSA's internationalization drive (see Chapter 8), affiliates abroad have absorbed increasingly large volumes of the nation's exports: in the late 1980s, their share rose to 40–50 per cent.

7.3 Pricing Mechanism

Venezuela is an unusual case among oil-producing nations because it is one of the few that still uses the method of notifying the price of its different crudes to its customers in advance. In other words, the price of its crudes – officially, at least – is not

Table 7.2: Venezuelan Crude Oil Term Contracts.

Customer	*Volume (1,000 b/d)*	*Destination*
Ruhr Öl	200	Germany
Citgo	130–170	USA
Champlin	130	USA
UnoVen	120	USA
Conoco	84.5	USA
Star Enterprise	42	USA
Sun	37	USA
Mobil	35	USA
Tarmac	30	UK
Coastal	26	USA
Seaview	25	USA
Nynäs	24	Germany
Shell	23	France
Exxon	20	USA
Amoco*	18	USA
Koch	15	USA
Texaco	15	USA
Chevron	12.5	USA
Trifinery	8.6	USA
Ergon	8	USA
Elf	6.5	France
Hunt	6	USA
Smith & Hollander	6	Europe
Exxon	5	Germany
Cepsa	3.5	Spain
Texaco	3	Germany
API	2	Italy
Lyondell	1.3	USA
Phibro	1.3	USA
Cameli	1	Italy

* This contract will be taken over by Citgo, as a result of the latter's purchase of Amoco's Savannah refinery, Amoco's sole facility supplied with Venezuelan crude.

Source: *PIW*, 27 April, 1992, Special supplement, p. 4.

indexed to that of any widely traded reference crude. However, the Venezuelan pricing method is not identical to the official price system used by oil-producing nations when OPEC's power was at its peak. Indeed the Venezuelan pricing system gets top marks for being opaque and confused (noteworthy characteristics, considering that the oil market is literally obsessed with the need for clear, transparent and straightforward price signalling).[9]

The system works in the following way. First, PdVSA notifies its term customers of the prices of the different crudes for the coming month. The customers are then expected to specify the volume of crude which they will be lifting (if any). The time limit for volume nomination is ten days before the beginning of that month (on this date, PdVSA's operating companies set the loading dates for each of the cargoes available on the month). After nominating volume, the customer is expected to nominate a date for 'price setting', a somewhat complicated operation whose purpose is to link the official posted price to the variations in the oil market at large. An arithmetic average (which we will call *A1*) of the prices of WTI, WTS and ANS (mean of *Platt's* quotations for these crudes) is computed, for the five days around the date selected by the customer (two days before, two after, and the date proper).[10] Obviously, since each customer can choose a different date, depending on his views of where the market is heading, the result of this is that two identical cargoes (same crude, same volume) can conceivably be lifted on the same date, but with different prices for each of them. After the cargo has been lifted, another average of the prices of the same crudes (which we will call *A2*) is calculated for the five days around the Bill of Lading date. After this step is completed, the final price of each individual cargo is determined, using the following formula:

Final price = Official price + (*A2* – *A1*)

Since the price of Venezuelan oil is determined casuistically, with reference to an official posted price which has been adjusted by a pricing formula of sorts, the pricing mechanism can best be characterized as a 'frame contract', semi-official pricing method.[11] It should be noted that exports to Venezuela's foreign ventures are not priced by this method. Rather, PdVSA invoices its subsidiaries abroad at an internal transfer price based on the feedstock's netback value. There are also persistent rumours that PdVSA has netback type deals with firms in which it does not have a shareholding interest.

7.4 Petroleum Products Exports

Until the mid-1940s, Venezuela exported only negligible

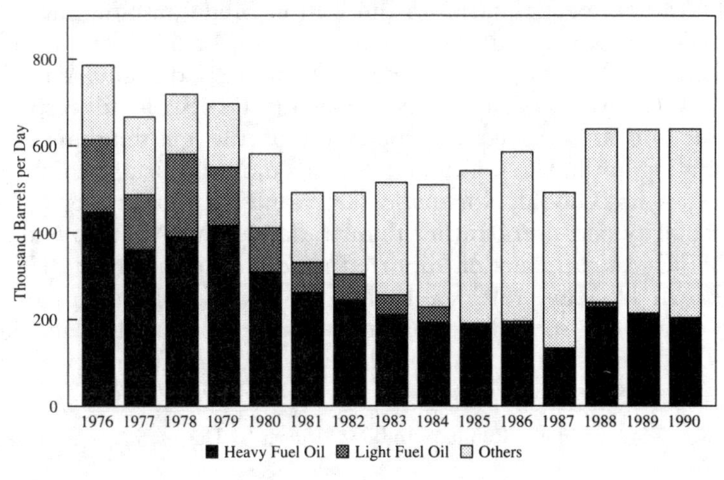

Source: PdVSA

Figure 7.2: Venezuelan Exports of Refined Products, 1976–90.

amounts of petroleum products. The promulgation of the 1943 Hydrocarbons Law, and the completion of the two refining complexes in the Paraguaná peninsula (Cardón and Amuay), brought about a complete reversal in this situation. From 1944 onwards, products exports became quite significant, and by 1970, accounted for about 30 per cent of the country's total petroleum exports. However, most of these exports (about 70 per cent) consisted of a low value product, fuel oil.

This unsatisfactory state of affairs continued for some time after the Venezuelan state took over the oil industry. However, in 1983, as a consequence of PdVSA's massive refinery upgrading programme, exports of light products began to rival those of fuel oil in volume. Throughout the 1980s, the relative share of fuel oil in the export slate continued to decline steadily, as Figure 7.2 shows. However, in terms of absolute volume, fuel oil exports are still quite significant.

The most important market for Venezuela's products exports is the USA (see Table 7.3). Indeed, PdVSA has become a very important player in many segments of the USA market for imported products. For instance, it is a key supplier of heating oil for the Boston Bingo forward cargo market, accounting for over 40 per cent of the imported product traded in it. Also, most of the traders participating in the Boston Bingo gasoline market used Venezuelan gasoline contracts to back their trades, although the rising cost of the contract has led many of them to other sources of supply. Nevertheless, PdVSA is still responsible for an important proportion of the gasoline imports arriving in the USA, although now its most important contract customers (Global, Northville, Getty and its affiliate Citgo) are primarily end-users, rather than traders. PdVSA is also one of the largest suppliers of imported residual fuel oil in the USA (*WPA*, 11 May, 1992, p. 4), and its affiliate Maraven caters to the USA specialty lubricants markets (in fact, it is the second biggest exporter of lube bases in the world after Exxon).[12]

Table 7.3: Venezuela. Refined Products Exports by Destination. Barrels per Day.

	1986	*1987*	*1988*	*1989*	*1990*
Total Exports excl. Bunkers	399,700	323,000	416,700	441,400	418,200
North America, of which	399,700	323,600	416,700	441,400	418,200
Canada	28,300	13,300	31,100	39,200	20,600
USA	371,400	310,200	385,700	402,200	397,600
Latin America	95,800	101,000	125,100	94,100	123,300
Western Europe	62,700	39,900	53,100	71,500	47,900
Africa	1,500	4,000	2,600	5,000	2,100
Asia and Far East	12,400	4,800	8,100	4,000	4,000
Unspecified	-	4,000	16,900	1,000	21,300
Total Exports excl. Bunkers	572,100	477,300	622,500	617,700	616,800
Bunkers	12,900	13,400	20,600	20,600	22,300

Source: OPEC, 1990, p. 92.

As far as the commercialization of its petroleum products output goes, PdVSA can no longer be regarded as residing in the backwaters of marketing, as it did in 1976. On the contrary, recent events prove that it has acquired considerable commercial prowess. Since 1990, for instance, its US affiliate Citgo (see Chapter 8) has waged a price war with the major oil companies in the USA, aimed at undermining their dominance in the distribution of jet fuel at key airports in the southern tier of the country. In so doing, it has aimed at a particularly sensitive spot in the oil majors' marketing operations because, due to the closely knit supply structures which hold sway in jet fuel markets, the relatively stable profits generated by jet sales, and the immense growth potential of jet markets, the majors have traditionally been very keen on keeping jet fuel distribution entirely to themselves.[13] However, airlines are finding Citgo's offers of long-term jet contracts, which undercut other suppliers' prices by 2–3 cents per gallon, very hard to resist. Thus, Citgo has been able to penetrate crucial airports, like Atlanta or Miami.[14] So far, the response of other oil companies to this expansion has been quite ineffective, limited either to limp pleadings to the airlines to buy domestic fuel, or to allegations of foul play (some of Citgo's competitors have said that its supply of Venezuelan crude gives it an unfair advantage). According to *Jet Fuel Intelligence*, one company has gone so far as to shut down operations in Atlanta and throughout Florida because of Citgo's aggressive price policies (4 May, 1992, pp. 1–2). Furthermore, the campaign has proved so successful that some airlines, overcoming both their wariness regarding Citgo's state-owned status and their reluctance to give offence to some of their traditional suppliers, have asked the company to establish a presence at key airports west of the Rocky Mountain divide (*JFI*, 23 September, 1991).

Citgo's strategy regarding the US gasoline market closely mirrors the one it has adopted for jet sales, and it may prove to be just as distasteful to its competitors. Taking advantage of the high profit margins resulting from the 1990 Gulf crisis, Citgo has moved in a previously uncharted direction in order to further expand the number of its retail outlets.[15] In essence, Citgo has tried to bring independent dealers within its fold by ensuring them a fixed profit margin. The pricing mechanism whereby this is accomplished links the cost of gasoline to dealers to the prevailing retail average price for their specific location, minus a net discount (equivalent to a gross margin for the dealer, of the order of about 10 cents a gallon).[16] This type of arrangement, reminiscent of the netback type pricing which pervaded the oil market during the early 1980s, is 'uncharacteristic of the USA retail motor fuel industry and could potentially revolutionize the gasoline distribution system' (*PIW*, 15 April, 1991, p. 2). Its results have been quite positive: in 1990, for instance, Citgo increased its volume of gasoline sales by 16 per cent, as compared to an overall USA gasoline market growth of 3 per cent.[17]

Nevertheless, for all its remarkable progress, PdVSA's marketing skills in some areas – notably trading – are still quite weak. Its domestic output of gasoline and distillates, for instance, is still sold mainly through traders who lift the products f.o.b. in Venezuela.[18] PdVSA, at the request of some of its customers, has explored the possibility of direct c.i.f. sales (which could give PdVSA additional profits), but nothing has come of this so far. Another, probably more important, problem in PdVSA's marketing approach is its summary dismissal of the use of futures contracts as an instrument to manage the risks inherent in its crude and products sales. This attitude can probably be traced to deeply entrenched corporate values which see any paper trading activities as speculative and risky or at best, marginal and irrelevant.[19] Thus, PdVSA does not trade oil in futures (except through Citgo and Uno-Ven),[20] and so far, has been markedly unwilling to employ OTC financial instruments, like options, warrants or swaps.

7.5 The Supply of the Domestic Venezuelan Market

Before 1964, the supply of petroleum products for the Venezuelan internal market was conducted mainly by the subsidiaries of three international oil companies: Creole, Shell de Venezuela and Mobil de Venezuela. In that year, however, the government issued a decree whose objective was to strengthen the role of CVP in the domestic market, and which eroded the position of the foreign multinationals. Decree 187 established that the participation of the state oil company in the market had to reach a level of 33 per cent by 1968. In subsequent years, this process whereby the government made parts of the Venezuelan market out of bounds for foreign companies continued; it reached its peak in 1973, with the promulgation of the Law Reserving the Exploitation of the Internal Market for Products Derived from Hydrocarbons to the State. This law established that the Venezuelan state, by means of its constitutional powers, would henceforth reserve to itself the activities of importing, transport, supply, storage, distribution and retailing of all fuels, LPG, natural gas, lubricants and lube bases, hydraulic systems fluids, brake fluids, petrolates, paraffins and asphalts (CEPET, 1989, v. II, p. 113).

The state's monopolization of these activities, which would have given CVP a 100 per cent participation in the Venezuelan internal market, was supposed to be completed in 1976. When the nationalization of the oil industry intervened, however, President Pérez decided to suspend any further expansion of CVP in the domestic market, at a moment when the company accounted for 68 per cent of the market. This left Maraven with a 19 per cent market share, and Lagoven and Llanoven with 11 and 2 per cent shares, respectively. In 1978, Meneven entered the market, and rapidly achieved a 5 per cent share. The absorption of Meneven by Corpoven (itself a fusion of CVP and Llanoven) left Corpoven as the operating subsidiary with the strongest presence in the internal market.

The size of the domestic products market, when compared to that of the export market for Venezuelan crudes and products, has always been quite small. However, over the last 30 years, this market has experienced a great deal of growth. Between 1960 and 1970, for instance, it grew at an annual rate of 6.2 per cent and between 1970 and 1975 its rate of growth averaged 9.3 per cent. Since then, its expansion has continued, although at a lesser pace: between 1976 and 1980, its growth rate averaged 8 per cent a year; for the 1980–5 period, it was only 0.8 per cent (a result of the great economic contraction which occurred at this time). Finally, between 1985 and 1990, PdVSA's sales in the domestic market expanded at an average yearly rate of 3.7 per cent (PdVSA, 1976–90; CEPET, 1989, v. II, p. 119). Table 7.4, shows the evolution of the internal market for petroleum products since 1976.

Table 7.4: PdVSA. Sales of Products in the Venezuelan Local Market, 1976–90. Barrels per Day.

Year	*LPG*	*Gasoline*	*Diesel*	*Fuel Oil*	*Others*
1976	18,000	115,000	46,000	31,000	34,000
1980	23,000	159,000	73,000	59,000	41,000
1985	28,000	163,000	53,000	58,000	35,000
1989	35,000	162,000	57,000	43,000	52,000
1990	39,000	164,000	61,000	34,000	32,000

Source: PdVSA

The 1973 internal market law defined the supply of the Venezuelan domestic products market as a service to be conducted in the public interest. This law, together with the 1975 nationalization law, made it the oil industry's duty to 'guarantee the supply of hydrocarbons and products derived thereof which the country needs for its development'. Furthermore, both laws established that, in addition, the oil industry would have to 'provide the exchequer with foreign exchange, and function as a direct stimulus to national development' (CEPET, 1989, v. II, p. 113).

Obviously, pricing policy would be crucial to determine the extent to which any of these goals were fulfilled. Therefore, the government set down some policy guidelines intended to direct PdVSA's pricing policy. These directives established that the prices for products would be set according to 'the energy value of the products, the export value of the products, the relative abundance of the products and the socioeconomic impact of their price' (ibid.). However, for much of the nationalized industry's history, the only criterion that has really carried any weight with Venezuelan politicians has been the fourth.[21] In other words, the prices of petroleum products in Venezuela have been kept low as a matter of government policy, in order to contribute to the legitimation of the state in the minds of its less well off, and middle class, citizens.

Up to 1981, according to PdVSA ex-president Rafael Alfonzo Ravard, because of the very low average selling price of refined products in the domestic Venezuelan market, the implicit price of the crude used to obtain these products was only about $5 per barrel. This price, equivalent to an implicit subsidy exceeding $100,000,000 per year (Evans, 1991, p. 390), severely restricted PdVSA's domestic sales revenues. It also contributed to a reduction in the availability of oil for export because it fostered indiscriminate consumption in Venezuela, and the appearance of large-scale gasoline smuggling operations (by means of road tankers) from Venezuela to Colombia.[22] When he presented these numbers to the government, General Ravard insisted that if this subsidy was to be maintained, the government had to take upon itself the onus of funding it directly, because its policy of making PdVSA pay indirectly would irreparably damage the financial base of the company (ibid.).

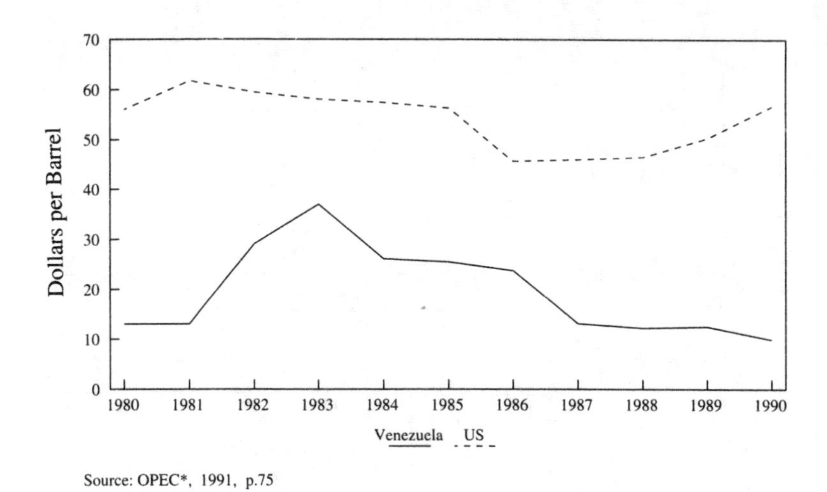

Source: OPEC*, 1991, p.75

Figure 7.3: Retail Prices for Premium Gasoline in Venezuela and the USA.

After 1982, as a result of the deterioration of the Venezuelan economy, the government decided to pay heed to Alfonzo Ravard's pleas. Thus, it tried to improve the financial basis of PdVSA's operations by bringing the domestic price of products closer to the prevailing world market prices.[23] In March 1982, therefore, the prices of gasoline (the product which had shown the largest increases in consumption rates) were substantially increased. However, due to the octane shortage which the country was undergoing at this point, the price increases for premium gasoline were far bigger than those for regular gasoline (112 per cent compared to 40 per cent; see Figures 7.3 and 7.4). These price increases, combined with the acute economic recession of the early 1980s, reduced Venezuela's gasoline consumption to an estimated 325,000 b/d by 1985, compared to a peak level of 390,000 b/d in 1982. After 1985, price increases continued, but these were solely linked to inflation: in dollar terms, the prices for both grades fell to less than half of what they had been in the years 1982–3, and the differential between international and domestic prices widened once again (see Figures 7.3 and 7.4).[24] Not surprisingly, PdVSA's losses from domestic fuel sales were quite substantial (ibid.).[25]

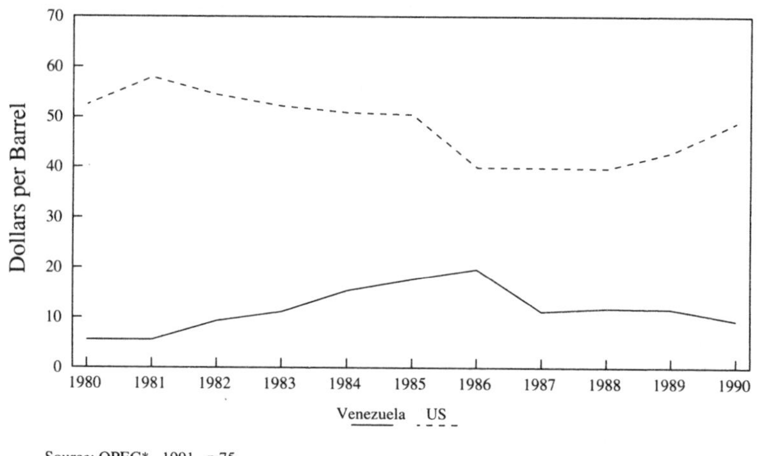

Source: OPEC*, 1991, p.75

Figure 7.4: Retail Prices of Regular Gasoline in Venezuela and the USA.

Until recently, the Venezuelan gasoline market was almost unique in Latin America because the price differential between the two grades of gasoline was enormous. In other parts of the continent, this did not happen because the cross price elasticity between the two grades was very high. Thus, substitution effects made the existence of sizeable and lasting price differentials almost impossible (Sterner, 1989, p. 39). In the early 1980s, Venezuela's situation was different because of the great octane scarcity in the country, itself a consequence of the oil industry's chronic lack of conversion capacity. However, as Figure 7.5 shows, the prices of both grades have been edging closer together of late. This can be said to be a logical consequence of the increasing complexity of the Venezuelan refining system.

In addition, the price of kerosene in Venezuela has not followed the general Latin American pattern. In most of the continent, kerosene is the most heavily subsidized product, because it is used for lighting and cooking in poor households (Sterner, 1989, p. 37). In Venezuela, however, kerosene has consistently been more expensive than diesel (see Figure 7.6); before 1973, it was even more expensive than gasoline (ibid., p. 39). Diesel (which in Venezuela is heavily, but indirectly,

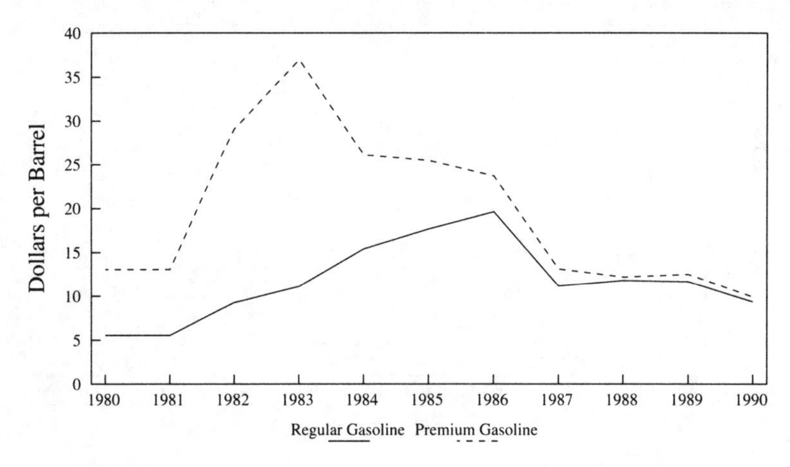

Figure 7.5: Retail Prices for Different Gasoline Grades in Venezuela.

used by the poor in the form of public transportation) has been the target of the largest implicit subsidies. Strangely enough, fuel oil, the one product which throughout history has been over-abundant in the country, is also the one product whose price in recent years can be said to bear even a passing

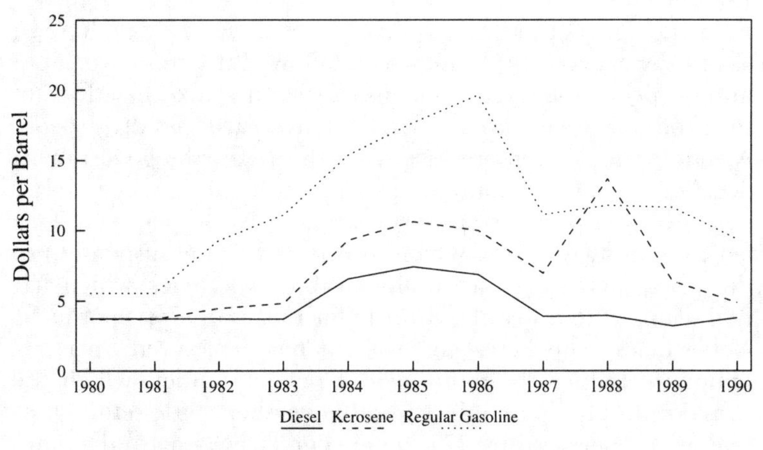

Figure 7.6: Retail Prices for Different Petroleum Products in Venezuela.

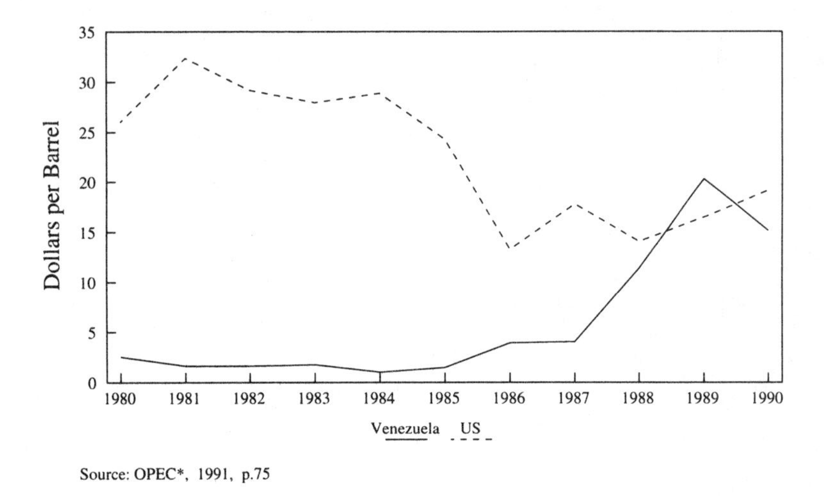

Figure 7.7: Wholesale Prices for Heavy Fuel Oil in Venezuela and the USA.

resemblance to international prices. The explanation for this apparent anomaly should be sought in the fact that the quotes for Venezuelan fuel oil prices are based on wholesale sales of Bunker C fuel sold mainly to foreign ships (and thus less subsidized than other products; see Sterner, 1989, p. 41, and Figures 7.6 and 7.7)

As the figures plotting gasoline prices show, the devaluation of the Venezuelan currency which followed the country's lapse into deeper economic troubles after 1986 negated the effects of the products price increases implemented earlier in the decade. At one point, under pressure from the IMF, the government decided to increase domestic fuel prices by an average of 101 per cent, and public transportation fares by 30 per cent. This led to nationwide riots, which in turn led to the suspension of the price increases – an unfavourable experience which left politicians with a healthy distaste for raising the price of fuels. Nevertheless, the Pérez government has persevered in its attempts to reduce the monumental revenue outlets which the subsidies imply. With effect from November 1991, retail gasoline prices increased by 27–29 per cent. According to the plan, monthly increases will more than double its price by 1992. The price of diesel and other products was also adjusted, but whether this will take domestic prices closer to export prices is uncertain at the moment, since a similar plan was aborted when it was already in full implementation stage in the early days of 1991.

Notes

1. Except Canada and Mexico, which were not considered in the programme. At this point in time, however, Mexico was a very marginal supplier of crude to the USA.
2. By 1970, the world oil price was $1.30 per barrel, and the US domestic price was $3.18 per barrel.
3. The maturity date for long-term loans has since been shortened to 12 years.
4. This is especially true of Mexico, who used the San José accord to highlight the independence of its foreign policy from that of the USA by supplying the revolutionary regime in Nicaragua with oil, at a moment when the Reagan administration was doing its utmost to force the Sandinistas out of power.
5. Mexico, for example, stopped sending oil to Nicaragua after the Sandinistas had run up a half billion dollar oil debt, and had given no indications of being able to pay any of this money back. Oil supplies to Haiti were interrupted for the same reasons. Mexico has also complained that the funds generated by the programme have not been used well. Ideally, the projects financed by the easy-credit terms were supposed to benefit Venezuelan and Mexican companies. Unfortunately, very few projects have materialized. Therefore, Mexico has suggested that the World Bank be brought in to monitor the way in which the beneficiary countries spend these funds. This does not mean, of course, that the political aspect of the accord has ceased to exist. Venezuela, for instance, has said that it would be willing to include Cuba in the accord, provided that democratic reforms in the island were enacted. Mexico, for its part, stopped oil supplies to Panama as a result of a quarrel with the US-imposed government of Guillermo Endara.
6. Of course, this does not mean that the contract is strictly adhered to *à outrance*; cargo-by-cargo transactions with regular customers are sometimes done, when PEMEX has crude available for supplemental sales above contractual commitments.
7. CVP's exports were always negligible.
8. The Venezuelans had learned how difficult it was for inexperienced players to place substantial volumes of crude oil in the international marketplace in the late 1950s, when the government elected to receive part of its royalty payments in the form of crude oil, to be resold at higher prices than those on which the operating companies were taxed. This exercise, however, proved to be anything but a commercial success.
9. PdVSA has also been known to use attractively priced products packages to 'sweeten' its crude deals.

10. If the customer does not nominate the price during the specified period, it will be the one corresponding to five days before the date of the Bill of Lading.
11. Frame contracts are those that price crude on a cargo-by-cargo basis (*PIW*, 27 April, 1992, Special supplement, p. 1). PdVSA has also been known to sell some of its crudes (notably Furrial) on an ANS based formula.
12. Most of Maraven's lube exports (8,000 b/d) are produced at the Curaçao refinery, while 2,000 b/d come from the lube trains at Cardón. Maraven has sometimes used Trinidad and Tobago's Trintoc refinery at Point-à-Pierre to produce lubes on its account.
13. For example, Texaco did not include jet fuel marketing in the Star Enterprise joint venture with Saudi Aramco.
14. Citgo's main success so far has been Atlanta International Airport, the biggest in the world in terms of the number of flights operating from it. The growing number of overseas flights originating there (a consequence of the construction of a new international passenger terminal at the airport and the acquisition of Pan Am's international routes by Delta), together with the designation of the Lake Charles refinery as a 'foreign-trade zone', and the decision of the airline corsortium operating in Atlanta to give foreign-trade zone fuel preference at the supply facilities for the new terminal, has enabled Citgo to build an unbeatable edge in supplying jet fuel to international carriers at this airport. According to *JFI*, 'other oil companies would have to reduce their domestic prices 4 per cent to compete with Citgo's tax-free status' (4 May, 1992, p.1). Citgo must pay the usual taxes and duties on petroleum products from the facility sold into the US market, of course.
15. Citgo is considered to be the fastest growing branded petroleum marketer in the USA. The number of Citgo's retail sites has jumped from 3,500 in 1985 to 9,800 by the beginning of 1991.
16. This incentive is in line with historical US gross retail margins, which have hovered between 10 and 15 cents (*PIW*, 15 April, 1991, p. 2).
17. Citgo, however, has decided to slow down gasoline sales growth to 8 per cent a year, and to emphasize the upgrading of its image. Thus, instead of adding thousands more jobber stations every year, the company will try to boost its sales at a smaller number of existing stations. It also plans to concentrate on premium gasoline sales (*U.S. Oil Week*, 21 September, 1992, p.5). It should also be mentioned that Citgo's image with gasoline retailers could not be better. In late 1992, for instance, a survey among members of the Petroleum Marketers Association of America ranked Citgo as the best supplier company overall, and gave it the top marks in 21 out of 25 categories, and 4 out of 6 major categories (*U.S. Oil Week*, 12 October, 1992, p.1).
18. Its sales of fuel oil to markets outside the USA are handled mainly by BP, with whom PdVSA signed a commercialization agreement.
19. A senior PdVSA official gave a very clear illustration of this when he asked: 'Why should we hedge for a few pennies a barrel?'(*EC*, 9 August, 1991).

20. In 1988, two of PdVSA's affiliates held a paper trading simulation in New York and Caracas, but the company's involvement in this type of activity has not gone beyond this.
21. As Coronel says, 'pricing policies . . . have almost entirely been dictated by political considerations' (1983, p. 144).
22. In 1980, for example, Colombia's gasoline prices were three times higher than Venezuela's.
23. Prior to March 1982, petroleum products prices in Venezuela were the lowest in the world, bar those of Saudi Arabia.
24. The prices of these products in the USA (a good substitute for world prices) are provided, to show the magnitude of the implicit Venezuelan subsidy.
25. For instance, from 1989 to 1992, the company's losses on its domestic market activities have been estimated at about $9.4 billion (*IPE*, 1990, p. 165).

8 PdVSA'S REFINING INTERESTS OVERSEAS

8.1 Internationalization and PdVSA's Commercial Strategy

The tight supply situation prevailing in the international oil market at the end of the 1970s gave PdVSA a respite from the tough commercial environment it had encountered when it came into being. Having had to yield to the demands of the Venezuelan ex-concessionaires during the early years of its existence, PdVSA then found itself negotiating, price-wise, from a position of great strength. However, the company did not let a false sense of security deceive it into believing that its oil would always 'sell itself', as it seemed to be doing at that moment. Thus, notwithstanding the exceptionally favourable circumstances under which it was operating, PdVSA submitted all of its commercial activities to a careful examination, and it reached the following conclusions (here summed up by A. Quirós, ex-president of Maraven):

> We noticed three things: first, that residuals [were] not our best product for future exports; second, that we had an abundance of heavy oils, and third, that these heavy oils should be dedicated to the making of special products such as lubricants or be subject to deep conversion in order to obtain marketable products (Coronel, 1983, p. 219).

Against this background, PdVSA designed a marketing strategy whose main objectives were the maximization of the level of government revenues coming from oil, and the minimization of fluctuations in these revenues (rather unsurprising objectives, considering the total dependence of the government budget on oil revenues). This strategy was to serve as a commercial guideline for its oil activities, and its main points were:

a) Preference of supply should always be given to final consumers, rather than traders.
b) Commercial activities of a speculative character should be avoided.

c) In terms of sales, term contracts should be the preferred commercial medium. Spot sales are to be used either to achieve revenue optimization under certain market circumstances, or to solve operational problems.
d) Excessive dependence on a small number of clients should be avoided.
e) Under equal economic conditions, and with PdVSA's commercial situation allowing, state-owned companies should be given preference of supply.
f) Sales of products or crude should be diversified according to end-uses and oriented towards those markets which either offer the highest economic value, or can be considered secure outlets.
g) The penetration of markets should be carried out in an orderly fashion, by means of identifying commercial opportunities which will not affect the price structure in that market.
h) Heavy crude sales to specialty or high conversion refiners should be given preference, in order to avoid increases in residual supply caused by sales to general purpose and/or less sophisticated, refineries.
i) Strategic clients for heavy crudes should be protected, if necessary, by means of preferential supply agreements.
j) Every effort should be made to channel sales of relatively less valuable products to the Venezuelan internal market (CEPET, 1989, v. II, p. 158).

Just as this strategy was leaving the drawing board, some Venezuelan decision-makers proposed a complementary element for it, which they considered would reduce the company's volumetric risk even more: the acquisition of downstream assets in important consumer countries. In 1981, a rapidly deteriorating market situation convinced PdVSA that it had to shift its commercial strategy into a higher gear, and that the way forward lay in pursuing an aggressive joint venture programme with refiners in important consumer countries. Two years later, the implementation of this so-called 'internationalization' programme began, with the signing of an agreement which gave PdVSA a 50 per cent share in Veba Öl's refining business. Since then, PdVSA's internationalization has become the best known aspect of Venezuelan oil policy outside the country: in addition to giving PdVSA an extremely high corporate profile in international circles, it has also transformed Venezuela into the OPEC member with the largest amount of refining capacity outside its borders.

Since 1983, PdVSA's acquisitions of assets abroad have been governed by a set of fairly well-defined policy directives. First of all, the company established that it would only acquire an interest in companies which have their own refining and marketing facilities. Secondly, it said that it would always insist on conditions which, on the one hand, would give PdVSA a reasonable degree of control over company strategy and, on the other hand, would enable it to place a part of Venezuela's production in the form of petroleum products.[1] Thirdly, PdVSA singled out as desirable partners those refiners who had the capability to process most of the wide range of crudes produced by Venezuela and/or whose infrastructure and economic conditions justified the emplacement of deep conversion plants in their refining facilities in the future. Last, but by no means least, the company said that it would always try to obtain financial terms which would permit it to acquire the assets by putting the lowest possible quantities of cash up front. Its preferred method to liquidate any debts incurred as a result of asset acquisitions would involve payment in kind (with crude supplies), or the proceeds from dividends or the cash flow generated by the target company, or a combination of both.

PDVSA's internationalization programme has provided the company with an assured destiny for 40–50 per cent of its exports.[2] The company's stated objective of placing 700,000 b/d of crude in overseas joint ventures (CEPET, 1989, v. II, p. 177) is currently met with ease, as Tables 8.1 and 8.2 demonstrate; Tables 8.3a and 8.3b show the main characteristics of the refineries in which Venezuela has a shareholding interest, or owns completely. In the following sections, we shall examine each of PdVSA's refining ventures overseas in more detail.

8.2 Ruhr Öl GmbH

PdVSA struck its first overseas downstream integration deal with Veba Öl (a subsidiary of Veba A.G., one of the biggest

Table 8.1: PdVSA. Foreign Refining Interests.

	Country	*Partners (if applicable)*	*PdVSA share (%)*	*Venezuelan crude supplied***
Citgo Petroleum Corp., Lake Charles, La.	USA	-	100	170,000 b/d
Corpus Christi, Tx.	USA	-	100	130,000 b/d
Paulsboro, N.J.	USA	-	100	30,000 b/d
Savannah, Ga.	USA	-	100	12,600 b/d
Uno-Ven Co., Lemont, Ill.	USA	Unocal	50	135,000 b/d
Refinería Isla S.A., Emmastad, Curaçao	Netherlands Antilles	-	Leased	180,000 b/d
Nynäs Petroleum NV, Antwerp	Belgium	Neste Öy	50	10,000 b/d
AB Nynäs Petroleum, Göteborg	Sweden	Neste Öy	50	12,000 b/d
Nynäshamn	Sweden	Neste Öy	50	5,000 b/d
Ruhr Öl GmbH, Gelsenkirchen	Germany	Veba Öl	50	
Oberrhenische Mineralölwerke GmbH, Karlsruhe	Germany	Veba Öl*	16.5	104,000 b/d +
Erdöl Raffinerie Neustadt GmbH, Neustadt am Donau	Germany	Veba Öl*	12.5	
PCK Schwedt AG, Schwedt	Germany	Veba Öl*	18.75	41,000 b/d
			Total:	829,600 b/d

+ Includes all German refineries except Schwedt.
* These refineries are controlled by other shareholding interests besides those of Veba and PdVSA.
** This denotes contractual volumes. Actual supplies, when averaged, may turn out to be less than this, because of the Venezuelan practice of only insisting that a percentage of a contract's volume be lifted in a year.
Source: *WPA*, 27 January, 1991, p. 18

Table 8.2: PdVSA. Level of Downstream Integration, Excluding Leases. Barrels per Day.

Domestic refining capacity	1,171,000
Foreign refining capacity***	902,410
Crude output (1991 *O&GJ* estimate)	2,341,300
Refining capacity as percentage of output	88.6
Supplies to own foreign ventures	649,600
As percentage of output	28
Additional earnings per downstream barrel (dollars)	
Export refineries*	3.93
Foreign ventures**	0.93

* *PIW* estimates of incremental refining margins for 1990–91, excluding capital costs.
** *PIW* estimates based on full operating profits for Citgo.
*** Considers only percentage participation in those refineries that PdVSA does not own completely.
Sources: *WPA*, 27 January, 1991, p. 18; *IPE*; *PIW*, 25 November, 1991.

consortia in the Federal Republic of Germany). This deal established a company known as Ruhr Öl, which became the owner of a refining complex near Gelsenkirchen, with a processing capacity of 250,000 b/d.[3] The PdVSA–Veba agreement gave both partners a 50 per cent interest in the new company,[4] and provided for equal representation on the board for the two companies. Veba was left in charge of Ruhr's day-to-day operations; likewise, the agreement specified that the marketing, distribution and sales of the refined products from Gelsenkirchen would be conducted through Veba subsidiaries (with the participation of management personnel from Venezuela). The Ruhr Öl agreement also established a research and development accord between Veba and INTEVEP, in order to investigate and evaluate techniques for the handling of heavy crudes and the high conversion of residuals or extra heavy crudes.

In 1985, the Ruhr Öl charter was expanded to cover the whole of Veba Öl's refining and primary petrochemical operations. In this way, Ruhr acquired a 25 per cent share in the Erdöl Raffinerie in Neustadt, a 100 per cent share in the petrochemical plant at Münchmünster in Bavaria and a 33 per cent share in the Oberrhenische Minaralölwerke refinery in Karlsruhe. This gave Ruhr an additional 85,200 b/d of refining

Table 8.3a: Venezuelan Refining Ventures in Europe. Charge Capacities. Barrels per Day.

Process	*Antwerp*	*Göteborg*	*Nynashämn*	*Gelsenkirchen*	*Karlsruhe*	*Neustadt*	*Schwedt*
Crude	15,000	12,500	28,000	215,400	174,000	144,000	230,000
Vacuum Distillation	12,500	9,000	31,000	48,200	102,000	65,400	94,300
Visbreaking	-	-	-	16,000	31,000	13,000	26,800
Delayed coking	-	-	-	28,000	-	-	-
FCC	-	-	-	-	66,000	26,000	27,000
Reforming	-	-	-	38,900	23,700	17,700	41,000
Hydro-cracking	-	-	-	30,000	-	-	-
Hydro-refining	-	-	-	63,000	96,000	27,200	64,500
Hydro-treating	-	-	2,500	93,000	37,400	22,500	93,710
Alkylation*	-	-	-	-	10,200	-	-
Arom/Isom*	-	-	-	15,190	-	-	12,000
Lubes*	-	-	2,500	-	-	-	-
Asphalt*	12,000	8,500	20,000	6,000	11,800	5,400	2,400
Hydrogen, mcf/d	-	-	1.9	25.0	-	-	8.1
Coke, tonnes/day	-	-	-	1,200	-	-	-

*Production capacity, b/cd
Source: *O&GJ*, 23 December, 1991.

Table 8.3b: Venezuelan Refining Ventures in America and the Caribbean. Charge Capacities. Barrels per Day. Includes Leases.

Process	*Paulsboro*	*Corpus Christi*	*Savannah*	*Lake Charles*	*Lemont*	*Curaçao*
Crude	80,000	132,500	28,000	320,000	147,000	320,000
Vacuum Distillation	35,000	80,000	-	75,000	58,000	158,000
Thermal cracking	-	-	-	-	-	67,000
Delayed coking	-	33,500	-	88,000	27,900	-
FCC	-	76,500	-150,000	58,000	42,000	
Reforming	-	54,000	-	106,000	29,800	15,000
Hydro-cracking	-	-	-	45,000	-	-
Hydro-refining	-	106,700	-	40,000	-	-
Hydro-treating	-	60,000	-	173,000	103,800	40,000
Alkylation*	-	19,200	-	23,000	18,000	5,000
Arom/Isom*	-	10,500	-	27,000	10,900	-
Lubes*	-	-	-	9,000	-	8,000
Asphalt*	-	-	22,500	-	3,600	8,000
Hydrogen, mcf/d	-	-	-	-	11.0	-
Coke, tonnes/day	-	1,950	-	4,000	2,000	-

*Production capacity, b/cd

Source: *O&GJ*, 1992.

capacity, as well as a 200,000 tonnes per annum ethylene production capacity and a 140,000 tonnes per year propylene production capacity. The 1985 deal also brought within the Ruhr fold other very valuable properties. In northern Europe, Ruhr acquired a 38 per cent interest in the Maarschap Europoort terminal in Rotterdam, a 20 per cent interest in the Rotterdam–Rhein pipeline, and a 25.1 per cent interest in both the Nordwest Ölleitung terminal in Wilhelmshaven and the pipeline connecting this port with the Ruhr region. In southern Europe, Ruhr acquired similar shareholding interests in marine terminals in the ports of Lavera (France) and Trieste (Italy), as well as in the pipelines (the Transalpine and South European) that connect these ports with its refineries. The Lavera and Trieste terminals also give Ruhr access to storage installations with a capacity of 37 mb, thus conferring on Venezuela a great degree of flexibility for its European operations. In March 1991, Ruhr Öl acquired 37.5 per cent of the 240,000 b/d Schwedt refinery, the largest of the refineries in the former German Democratic Republic, even though its very dispersed ownership structure makes it difficult to arrange long-term supply contracts (like those beloved of PdVSA).

The distribution and sales of Ruhr Öl's products is done in a variety of ways. Motor fuels and lubricants are sold through the service stations of the Aral chain (in which Veba has a 56 per cent interest) in Germany, Belgium, Holland, Luxembourg, France, Switzerland, Austria and Eastern Europe, while gasoil, fuel oil and asphalt are marketed – mainly in Germany – by the Raab Karcher company (99.5 per cent Veba interest). By contrast, petrochemicals, fuel oil and aviation fuels are sold directly by Veba. Jet fuel is also marketed through the Aviation Fuel Services company, a joint venture between Veba and Lufthansa in which the former holds a 50 per cent shareholding interest.

PdVSA's decision to invest in West Germany was widely criticized in Venezuela at the time it was announced, especially on account of its cost ($200 to 250 million in 1983, plus $55 million in 1985). The move generated further political furore when it became known that 'the effective netbacks on crude refined under the [Ruhr] agreement were lower than Venezuela's official export prices for the same grades' (Evans, 1991, p. 390).

The Ruhr Öl controversy, however, did not deter PdVSA from pursuing other joint ventures, even bigger and more expensive than its partnership with Veba.

8.3 Citgo Petroleum Corporation

At the end of 1986, PdVSA and the Southland Corporation signed an agreement whereby the former would become the owner of 50 per cent of Citgo Petroleum Corporation, a Southland subsidiary. The deal, worth about $290 million, was linked with a 130,000 b/d supply contract to the Citgo Lake Charles (Louisiana) refinery. Then, in 1989, Southland announced that it was considering ways to convert its Citgo interest into cash, in order to pay the very large debt it had incurred in 1987, when it effected a repurchase of its stock. To avoid finding itself saddled with undesirable partners, PdVSA purchased Southland's remaining interest in the company, for the sum of $675 million. In 1990, PdVSA decided to restructure its USA operations, merging the assets of its Champlin and Seaview ventures with those of Citgo.

Citgo's Lake Charles 320,000 b/d refinery has an exceptionally high conversion index (its yields are 70 per cent gasoline and 16 per cent light distillates), which makes for a residual fuel oil output of just about 10,000 b/d. Its facilities include a 9,300 b/d lubricants plant, jointly owned with Conoco,[5] and a 2,400 b/d MTBE plant. The refinery also produces 100,000 tonnes per year of propylene, 100,000 tonnes per year of sulphur, 3.5 million barrels per year of propylene and 1 million tonnes per year of coke. Besides Venezuelan crudes, the refinery runs Mexican crude, and it continues to purchase crude from USA offshore leases. An agreement whereby Citgo purchases the whole of Occidental's USA crude production will remain in force until 1998. Citgo has to satisfy a gasoline demand superior to its own production capacity; therefore, to cover its commitments, it has to import gasoline from Venezuela, or to buy it in the USA market.

Citgo's distribution system is quite extensive. It has a shareholding participation in the Colonial[6] and Explorer pipeline systems, which link the USA Gulf Coast with the Great Lakes region and New England. It owns, partially or totally, more than 40 distribution plants, with a combined storage capacity of 18 mb of products. Citgo has about 10,000 retail outlets carrying its own brand name gasoline.

8.4 Champlin Refining Company

During the first quarter of 1987, PdVSA and the Union Pacific Corporation created a new company, called the Champlin Refining Company, in which each of the partners agreed to take a 50 per cent shareholding interest. Champlin Refining became the owner of a 165,000 b/d capacity refinery (along with its associated petrochemical plants) located in Corpus Christi, Texas. In 1988, by exercising a contractual option which stipulated that it could purchase Union Pacific's share in Champlin between 1989 and 1993, PdVSA became the sole owner of the company. The original 1987 PdVSA–Union Pacific agreement was concurrent with a crude supply contract for 80,000 b/d of Venezuelan crude. After PdVSA became the sole owner of Champlin, the supply contract was expanded to 130,000 b/d of crude and 10,000 b/d of naphtha, with PdVSA having the option to provide the whole of the refinery's feedstock if it so desires.

The Champlin refinery is a complex with plants located on both banks of the Nueces river, connected by means of crude and products pipelines.[7] It is a high conversion plant,[8] and its petrochemical installations can produce 2,700 b/d of cyclohexane, 2,591,000 tonnes per year of cumene, 5,000 b/d of benzene and toluene, 80 tonnes per day of sulphur and 1,600 b/d of MTBE. The market for Champlin's production is in the south-eastern United States; Louisiana, Texas, Florida, Virginia and Mississippi account for around 70 per cent of the company's sales. Champlin distributes and markets some 200,000 b/d of oil products through independent terminals under its own brand (in contrast to Citgo, it does not own a gas station chain).

8.5 AB Nynäs Petroleum

In June 1986, PdVSA acquired half of AB Nynäs, a unit of the Swedish conglomerate Axel Johnson,[9] for the sum of 165 million Swedish kronor ($23.5 million). The aim of this purchase was to consolidate the position of Venezuelan heavy crudes in the European markets for asphalt and naphthenic lubricants. Nynäs, one of Europe's largest asphalt and lubes producers and a company with a long experience in heavy crude refining, was a logical target in Venezuela's drive to expand sales of its heavy crudes in Europe. The deal gave PdVSA access to three specialized refineries, located in Göteborg, Nynäshamn and Antwerp (Belgium), with a joint capacity of 55,000 b/d.[10]

Nynäs produces about 500 different lube products, and accounts for about 18 per cent of the European market for naphthenic lubes. It has 19 sales offices and 16 distribution terminals for asphalt and lubes across Europe (Finland, Sweden, Norway, Denmark, England, Scotland, Belgium, Germany, France, Italy and Spain), along with a road tanker fleet and an asphalt dedicated ship and barge fleet.

8.6 The Uno-Ven Corporation

At the end of 1988, PdVSA signed a joint venture accord with Union Oil Corporation of California (Unocal) to take a 50 per cent share in Unocal's 151,000 b/d Lemont (Illinois) refinery (as well as in 12 fuel distribution terminals located in the states of Illinois, Michigan, Iowa, Ohio and Wisconsin; one aviation fuel terminal; 131 Unocal-owned service stations; a lube blending and packaging plant, located in Cincinnati, Ohio; and a venture which produces and sells super-premium grade petroleum coke), in a deal worth about $500 million (*O&GJ*, 12 December, 1988, p. 28). Under the terms of the agreement – which also gave PdVSA access to 3,300 independently owned service stations operating under the Unocal 76 trademark – PdVSA agreed to supply Uno-Ven with 135,000 b/d of light crude (with the balance of the refinery's diet, approximately 30,000 b/d, being made up of 26° API Canadian crude), in order to make and sell Unocal 76 branded products. Unfortunately, the logistics of supplying this type of crude to the Chicago area, from the USA Gulf Coast via the Capline pipeline, have resulted in the venture being less profitable than originally expected by PdVSA (*EC*, 9 August, 1991). Persistent rumours have circulated for a while that PdVSA was planning to buy Unocal's remaining share in Uno-Ven (the acquisition was supposed to take place in the second half of 1992).[11] However, the conjunction of Venezuela's economic woes and the venture's marginal economics has forced the PdVSA board to consider the sale of its stake in Uno-Ven, allegedly to the Kuwaiti national oil company.[12] (*PON*, 1 September, 1992, p.3). Thus, it is logical to expect that sometime in the future, PdVSA will try to divest itself of what some have called 'the poor sister' in the company's investments in the USA.

8.7 Seaview Petroleum Company

In late 1990, Citgo signed a general partnership agreement with the Seaview Petroleum Company, under the terms of which it became the holder of 50 per cent of Seaview's asphalt and refining business. Seaview's 84,000 b/d asphalt refinery and marine terminal in Paulsboro, New Jersey, provided a toehold for PdVSA to break into the very lucrative USA East Coast market. In March 1991, Citgo acquired the remaining 50 per cent of Seaview. Although the Paulsboro refinery is primarily dedicated to asphalt production, it is a logical candidate for upgrading.

8.8 Citgo Petroleum Corporation, Savannah

On 22 June, 1992, it was announced that Citgo would purchase Amoco Oil Company's 28,000 b/d asphalt refinery in Savannah, and take over Amoco's asphalt refining and marketing operations. Also included in the sale were two distribution terminals (one in Chattanooga, Tennessee, and the other in Wilmington, North Carolina) and 135 railroad tank cars. Before Citgo purchased it, this refinery was running Boscán crude exclusively; therefore, it seems likely that PdVSA bought the facility because Amoco was considering closing it down, due to its corporate restructuring plans.[13]

8.9 Refinería Isla (Curazao), S.A.

In October 1985, a PdVSA subsidiary company, Refinería Isla, began start-up operations at a 320,000 b/d export refinery located in Emmastad, Curaçao. This refinery, one of the Caribbean's most complex plants, had ended up in the hands of the Netherlands Antilles' government because Royal Dutch/Shell (its original owner) had found it impossible to operate it at a profit, due to the worldwide glut in refining capacity which hung over the oil market during the first half of the 1980s as well as the increasing redundancy of the Caribbean as a major refining and distribution hub.[14] PdVSA leased the refinery for an initial period of five years, for a fee amounting to $11 million per year,[15] with the lease agreement being open for renewal every two years.[16] The leasing arrangement was particularly attractive for PdVSA because, in addition to the refinery, it included the use of the Curaçao Oil Terminal, built in 1974 and able to handle tankers of up to 550,000 dwt at six jetties. These marine installations, plus a 15 mb crude oil storage capacity, a 1 mb heated capacity, 750,000 barrels of segregated clean oil storage capacity, 1.5 mb of segregated heated fuel oil storage capacity and ample blending facilities, make Curaçao the largest – and probably the most versatile – supply and trading facility in the Caribbean.

PdVSA's original plans for the Curaçao refinery were to have its operating subsidiaries supplying it with approximately 150,000 b/d of light crude for processing under contract. The products resulting from this arrangement would contribute to Venezuela's export slate, with Corpoven, Lagoven and Maraven being responsible for their international marketing. The operations in Curaçao, however, seem to have exceeded expectations, because crude throughput in the refinery in the last few years has averaged 180,000 b/d. The main markets for Curaçao's products are Latin America and the USA (each accounts for about 35 per cent of sales). Only 15 per cent of the products go to Europe, while Venezuela and other destinations account for the remaining 14 per cent.[17]

8.10 Briggs Oil Ltd, Eastham Refinery Ltd, Lyondell Petrochemical Company: PdVSA's Latest Refining Acquisitions.

In August 1992, it was announced that AB Nynäs would acquire the refining assets of UK refiner Briggs Oil Ltd. This deal will give Nynäs 100 per cent ownership of a 10,000 b/d refinery in Dundee, Scotland, and a 50 per cent ownership of a 12,000 b/d refinery in Eastham, England. The acquisition will take Nynäs' share of the European asphalt market to 17 per cent (making it the second largest asphalt refiner in the region) and it will also permit PdVSA to place incremental volumes of Pilón and BCF-13 crudes in Europe.

PdVSA began a far more important deal in July 1992, when it was announced that Citgo would help Houston-based Lyondell Petrochemical company to raise its conversion capacity by 80,000 b/d, and in the process acquire a 50 per cent shareholding interest in Lyondell's deep conversion refinery.[18] This arrangement, will permit Venezuela to cover the whole of Lyondell's heavy crude requirements (200,000 b/d after the modifications).

Even before the final terms of the venture had been agreed upon by the partners, Lyondell began to phase out its supply contracts for heavy crudes with PEMEX and Saudi Aramco, and to step up its purchases of Venezuelan volumes.[19] The final contract for the joint venture is expected to be signed in the first half of 1993. Tables 8.4 and 8.5 show the characteristics of the latest Venezuelan refining joint ventures overseas.

Table 8.4: PdVSA Most Recent Downstream Acquisitions Overseas.

	Country	*Partners (if applicable)*	*PdVSA Share (%)*
Lyondell Petrochemicals Co. Houston, Texas	US	Lyondell	50*
Briggs Oil Ltd. Dundee	UK	-	100
Eastham Refinery Ltd. Eastham	UK	Shell	50

*PdVSA's shareholding will gradually go up to this level.
Source: O&GJ, 3 August, 1992, p. 24.

8.11 Examining the Case for PdVSA's Overseas Integration

The movement by the national oil companies of producing countries into the refining and marketing sectors of the main oil-consuming countries is a relatively recent development in

Table 8.5: Charge Capacities of Recent Venezuelan Refining Acquisitions. Barrels per Day.

Process	*Lyondell*	*Eastham*	*Dundee*
Crude	265,000	12,000	10,240
Vacuum Distillation	129,000	10,500	9,600
Delayed Coking	40,000	-	-
FCC	90,000	-	-
Reforming	110,000	-	-
Hydro-refining	129,000	-	-
Hydro-treating	170,000		
Alkylation*	14,000	-	-
Arom/Isom*	11,000	-	-
Lubes*	6,000	-	-
Asphalt*	-	8,000	6,950
Coke, tonnes per day	2,650	-	-

* Production capacity, b/cd
Source: O&GJ, 23 December, 1991

the history of the world oil market. PdVSA (which together with the Kuwait Petroleum Company has been at the forefront of this process) has used four main arguments to justify its investments abroad. The first says that integration favours revenue maximization through participation in all the segments of the industry, and a full utilization of assets. This argument, although plausible from an individual company's perspective at a particular point in time, does not hold when viewed from a long-term market-wide perspective. Revenue maximization is not a natural consequence of integration, because the gains coming from participation in all industry segments can only be maintained for a comparatively short period (the time it takes other market participants to realize that there are gains to be had from following this course of action, and to start acting accordingly). Undoubtedly, if a large number of market participants adopt this approach, success through integration will become progressively harder to achieve (*PII*, September 1988, p. 24). Therefore, the correct recipe for revenue maximization through asset management, should provide not only for integration, but also for divestiture: 'an integrated company, if it is to fully exploit the industry cycle to its best advantage, must be able and willing to alter the composition of its investment portfolio in a counter-cyclical manner' (*PII* September 1988, p. 31).

According to the second argument usually put forward in favour of downstream integration, a downstream network of refining and marketing assets may permit 'integrated' oil-producing countries to avoid the price-cutting wars which have generally accompanied cyclical contractions in the oil market, without having to cut their production volumes at the same time.[20] Unfortunately, this line of reasoning overlooks an intuition so obvious that it borders on the tautological; namely, that in a market where a crude over-supply situation exists, refined product demand is insufficient to allow everybody to produce at the level they wish, without concomitant price reductions. Granted, a producer that owns a refining system in a consuming country can, *strictu sensu*, assure itself that its crude production will be refined (because it can, after all, refuse to process crudes other than its own). But, in an oversupplied market, there is no way in which an integrated producer will be able to sell the products obtained from refining this crude if they are out of line with the price level found in that market. Or, to put it another way, if it switches from other crudes to run its own crude (overpriced from a market point of view), it will be incurring an opportunity cost; namely, the cost implied by foregoing the higher refining margins which running the other crude would have provided. Hence, an integrated producer, even if running its own oil through its own refining system, in the end has to face the same choices that confront non-integrated producers competing in a market: that is, if there is an over-supply of crude, whether to cut volume or to cut prices.[21] If neither are changed, it will find itself unable to sell the refined products (because they are too expensive), or achieving negative refining margins (because it sells the products at the same price as everybody else, but pays more for its feedstock), or running more attractively priced crudes through its system. The conclusion that downstream integration allows producing countries to maintain crude prices at a level which please them, while also maintaining their production volume, would hold if and only if such integrated producers were able to prevent other producers from gaining access to any given refinery, *regardless of the price at which they offered their crude.* This ability would, of course, give the integrated producers the power to force other producers to shut in capacity, but it presupposes a level of control by these producers over the downstream sectors of consuming countries which they do not have (and the Seven Sisters did).

The experiences of the various Venezuelan joint ventures confirm these intuitions. All of them are based on a netback-type crude supply arrangement; in this type of arrangement, the price of the crude supplies is set at whatever level is necessary to persuade the partner to refine the crude. In the cases where PdVSA enjoys a full equity ownership of the refineries, similar discounting (in transfer prices) can be hidden in the consolidated income statements of the company. In other words, as far as this argument goes, the benefits to be derived from downstream integration have been very much overstated.

The third common justification for downstream integration is revenue stabilization. This argument says that 'since crude oil prices define the unit cost of feedstocks at the refinery gate, and ultimately, the product costs to distributors and marketers . . . changes in crude prices will be smoothed by corresponding, opposite changes in costs downstream' (*PII*, September 1988, p. 35). PdVSA endorses this theory, as the following lines reveal:

> The natural to and fro movement of the market oscillates between a position of strength for the seller to positions where the buyers are supreme. Vertical integration guarantees that, by controlling all the segments of the industry, volume can be maintained and revenues maximized, either from the production segment or from refining and retail sales, according to the conditions of the market (CEPET, 1989, v. II, p. 175).

If this argument were true, vertical integration would make a lot of sense for a country like Venezuela, whose dependence on export earnings from oil is very high. 'Smoothing' her earnings function could be very important because export earnings instability can have adverse effects on economic growth. The uncertainty induced by fluctuations in export earnings may cause general instability in the economy because even though a change in total value of exports will affect directly the income of producers in the export sector (in this case, PdVSA), the variation in exports can also be transmitted – through a multiplier effect – to the other sectors of the economy. This can lead to changes in the domestic expenditure for consumption and investment (Obadan, 1986, *passim.*).

On a conceptual level, this third justification for integration seems to be well founded. After all, when oil prices fall, one can expect that profit on the downstream will compensate an integrated market player's losses on the upstream, and vice versa. However, the truth is that the gains derived from being integrated downstream can only be of a short-term character, because competition will inevitably bring about adjustments in refinery margins, taking these to levels in line with the long-term price perspectives of crude oil. A downstream network will only shield an exposed oil producer for a limited period, because high refining margins will act as a signal for more refining capacity to come on stream, and this will eventually bring the margins down.

Lastly, there are those who say that owning a high conversion refinery overseas gives PdVSA a secure outlet for certain streams of crude oil whose viscosity and sulphur and metal contents render them an unattractive option for many refiners. However, as Figure 8.1 shows, PdVSA's biggest refineries in the USA run, in the main, some of the lightest Venezuelan streams. Only the asphalt plants consistently process material with gravity inferior to 15° API.[22]

So what then is the point of having such a big overseas refining empire? Why buy a refinery and other downstream assets when the same results (say, the smoothing of a producer country's oil export earnings in periods of price volatility) can be obtained in with other mechanisms that do not entail the massive capital expenditures associated with refinery purchases. Mexico's successful use of the futures options market to hedge the equivalent of three months of its crude production in January 1991, is an obvious example of revenue stabilization through means other than refinery purchases. Other examples come to mind: the case of Iraq before its 1990 invasion of Kuwait shows how a country can increase its share of the USA crude market by resorting to attractive discounts and netback deals, the so-called 'poor man's integration'; while Nigeria

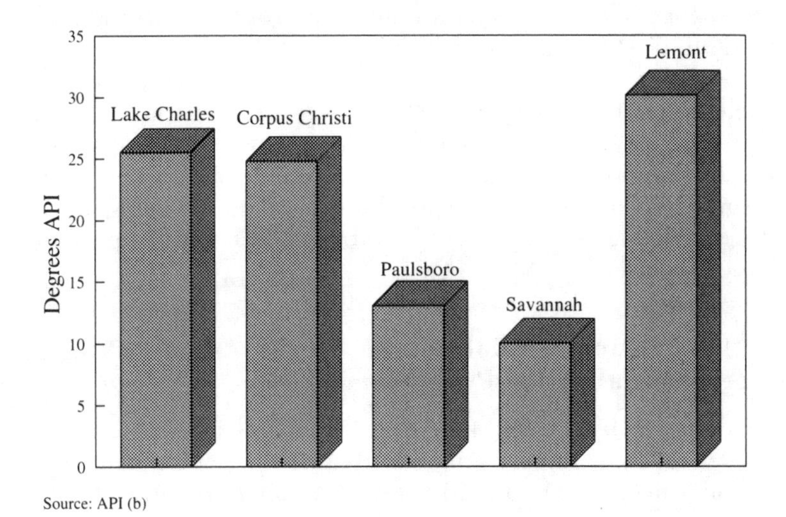

Source: API (b)

Figure 8.1: Average Gravity of Venezuelan Crude Processed by PdVSA's USA Refineries.

illustrates a producer that uses quasi-netback arrangements to maintain its sales volume constant.

However, if we accept as a given that it is worth PdVSA's while to be vertically integrated to a large degree (in other words, if we assume, as PdVSA does, that expanding its refining capacity, whether domestically or abroad, is *good*, and a *must*), internationalization is clearly a superior alternative to building export refineries (the other method of achieving a higher integration index). In PdVSA's case, this is not because the operating economics of its overseas refineries are better than those of its export refineries, due to the lower transportation cost of crudes, versus that of products.[23] PdVSA's preference for downstream acquisitions stems from the fact that the cost of these acquisitions can be covered by payments in kind. Thus, overseas acquisitions represent a viable way for PdVSA to add refining capacity, while getting around its grave handicap of having a very low cash ratio (itself a consequence of the way the government taxes the company). Building a refinery in Venezuela, needless to say, does not offer this option.[24]

Finally, there exists another strong justification behind the Venezuelan policy of overseas integration: the country's bad experience with the US mandatory import quotas, introduced in the late 1950s, and which did a great deal of harm to the Venezuelan oil industry. All of the country's attempts to get preferential treatment from the US government failed, and this left a lasting impression. Thus, in this context, Venezuela's acquisition of downstream assets in the USA can be seen as an insurance policy against the possibility of oil quotas being introduced once again in what still is, by far, her most important export market.[25]

8.12 Expansion or Contraction? The Future of PdVSA's Internationalization Programme

In January 1992, Celestino Armas (then Venezuela's minister for energy and mines) announced that the government was considering parting with 50 per cent of Citgo, because 'Venezuela's international integration policy had always stressed that PdVSA should [always] be in a 50–50 partnership with major oil companies in its foreign ventures' (*PON*, 14 January, 1992, p. 1). This explanation of the government's policy was rejected by PdVSA, which issued a statement that read: 'Citgo has been a fundamental factor to the growth of PDVSA, and has contributed substantially in improving the strategic positioning of this company . . . Whole or partial sale of Citgo, therefore, is not in PdVSA's current plans' (*PON*, 14 January, 1992, p. 4). Armas' statement did not convince public opinion either; most Venezuelans came to the conclusion that Citgo's proposed sale was an attempt by the government to raise large sums of cash fast, in order to alleviate its financial woes, and that it had nothing to do with policy directives restricting PdVSA to '50 per cent ownership in its foreign ventures'. Armas, seeing that his suggestion had drawn heavy flak from a variety of fronts, toned it down a little, saying that the sale of Citgo would only be considered after the US recession had definitely ended. After this placatory statement, the issue seemed to vanish from the agenda.

It resurfaced in June 1992, when the emir of Kuwait paid an official visit to the president of Venezuela. During the course of this visit, it was alleged that the emir offered $3 billion for a 50 per cent share in Citgo ($1.5 billion more than what the government thought it could get in January) and, in so doing, plunged Venezuelan oil policy into yet another morass of confusion.[26] On one hand, PdVSA said that 'no official offer had been made to [it]', but it recognized that it could not rule out 'the possibility that a deal [was being] discussed by the [two] heads of state' (*PON*, 12 June, 1992, p. 1). PdVSA also said that 'it would prefer a US partner, if at all, in the event a decision is made to sell half of Citgo' (ibid.). For its part, the Venezuelan government hinted that Kuwait's offer, though tempting, would not be taken up – not because the government had reversed its stance on the sale, but rather, that it was unwilling to face the inevitable political fallout if it ordered PdVSA to sell half of its most successful enterprise to a partner like Kuwait (with whom serious conflicts of interest would probably arise), and then used the proceeds to ease the federal deficit, at a time when the Venezuelan government was facing serious domestic difficulties. Such an action would have flagrantly contradicted the government's stated policy on sales of assets, which has always been that 'any proceeds from the sale of any part of Citgo would be reinvested internationally downstream' (*PON*, 12 June, 1992, p. 1). However, the tone and content of the government's statements and disclaimers did nothing to answer the question that sprang to many people's minds as a result of this *imbroglio*: if a more acceptable partner were to come along (say, a US major like Conoco), would the Venezuelan government still consider selling half of its star possession, the third largest petroleum retailer in the USA?[27]

This episode reveals the existence of a marked division within Venezuelan political circles regarding the worth of overseas downstream integration.[28] This division is especially evident when one considers that, just four days before the alleged Kuwaiti offer, *Platt's* announced that PdVSA was considering the acquisition of two of Shell's refineries in Great Britain, Lyondell's deep conversion refinery in Houston and Amoco's asphalt plant in Savannah (*PON*, 8 June, 1992, p. 1), as well as the buy-out of Unocal, its partner in Uno-Ven![29] Bearing this in mind, one can conclude that the future of PdVSA's internationalization process is surrounded with uncertainty and ambiguities. Whether the process will continue, and under what form are questions for which, at the moment, no clear answer exists. However, as the Amoco purchase demonstrated, it is likely that the one part in PdVSA's foreign ventures programme that might go ahead without big changes is its acquisition of new refineries. The fact that the government does not discourage these investments (because, since they are never paid for in cash, they do not absorb money that could be spent meeting social or infrastructure needs in Venezuela[30]) gives PdVSA an implicit incentive to go on with its foreign acquisitions. Thus, it is not difficult to understand how, even in the midst of the budgetary crisis affecting PdVSA, Citgo seemed to be engineering the takeover of the whole of the Uno-Ven Corporation, while making advances to Shell in order to buy two refineries in Britain, and to Lyondell Petrochemical Company in order to buy half of its 265,000 b/d refinery in Houston, Texas.[31]

However, the future of the upgrading plans which PdVSA had in mind for its foreign ventures – now that its budget has been axed, and its very ownership over the whole of Citgo questioned – is much more uncertain. In any case, if any of these projects eventually go ahead, it is logical to envisage the time horizon for their completion stretching beyond 1996 (PdVSA's original goal). Before the Venezuelan government's plans to sell half of Citgo cropped up, PdVSA planned to invest about $1.7 billion over the period 1992–7 to expand and upgrade the company's US refineries, in order to be better placed to comply with the Clean Air Act Amendment provisions, and also to refine larger volumes of heavier crudes (through these projects, PdVSA hoped that Citgo would be refining up to 220,000 b/d of 17° API crude[32] by 1996). About 75 per cent of PdVSA's investments in the USA was to have been absorbed by the Lake Charles plant, which would have received, among other things, a 70,000 b/d hydrotreater and its associated hydrogen plant. A 50,000 b/d expansion in the refinery's throughput capacity, costing an additional $300 million, was also in the offing. Citgo's Corpus Christi refinery was to have received $480 million, to pay for a new 4,500 b/d MTBE unit, a 2,200 b/d TAME unit, improvements for its waste water treatment system, and LPG storage systems.

PdVSA's plans for its refining ventures in Europe and

Curaçao were less ambitious than those proposed for its USA system. The company earmarked $60 million for the upgrading of the Curaçao refinery, while in Europe, it expected to invest about $800 million over the next three years in the Schwedt refinery, in order to make it capable of processing 180,000 b/d of Venezuelan crude. It was also considering the expansion of its retail outlet network in Europe (preferably through the acquisition of an existing network), and of Nynäs' operations in the Baltic. Regrettably, PdVSA's lack of liquid financial resources means that some of these projects may have to be abandoned.[33]

Some Venezuelan officials have suggested that the best way for PdVSA to get around its capital limitations is to enter into a partnership with a major international oil company (the names of Mobil and Conoco have cropped up in connection with this proposal), modelled after Texaco–Chevron's very successful Caltex joint venture. The idea behind this would be to exchange half of Citgo's shares in return for a similar amount of shares from the partner company. In the eyes of Venezuelan politicians, this proposal is attractive because it would tie the expansion of PdVSA's American operations to the fortunes of a US major, and this would render PdVSA's market position secure against any shift in US policy regarding investments by state-owned firms, and the like. PdVSA, however, has accorded this proposal only a lukewarm reception, arguing that US statutes adequately guarantee the safety of its investments in the country.[34] Its opposition would not be the prime obstacle in the path of this proposal, however. Much more difficult to overcome could be the natural reluctance of a publicly-owned US company to set up a very long-term partnership in the USA with PdVSA, unless the latter's relationship with the Venezuelan government were to change significantly, and all doubts about the country's political stability were definitely dispelled.[35]

Notes

1. In general, the company considers it indispensable that its prospective targets have means of channelling the products to the final consumer (i.e. a retail chain). Of late, this aspect has been given less importance

– at least in the USA – because Citgo's vast retailing network can be used to sell products from refiners that lack retail outlets (such as Lyondell, a rumoured PdVSA target).

2. This percentage refers to PdVSA's export levels up to the second quarter of 1990, when the company increased its production as a result of the 1991 Gulf crisis.
3. The complex is constituted by the Scholven and Horst refineries. These plants, located within 10 km. of each other, are interconnected by a network of oil and products pipelines which, in practice, permits their being operated as a single unit. The Gelsenkirchen complex has a good number of conversion units, as well as adjacent petrochemical plants for the production of ethylene, propylene, benzene, toluene, xylenes, ammonia and methanol.
4. This gave PdVSA the right to use up to half of Ruhr's refining capacity to run Venezuelan crudes. However, until the collapse of the Soviet Union, approximately one-fifth of the crude processed on PdVSA's account by Ruhr was Soviet Urals blend. This arrangement was the result of a bilateral co-operation agreement signed in November 1976, which stipulated that the Soviet Union would supply Venezuela's European customers with Urals crude in return for similar volumes of similar quality Venezuelan crude delivered to Cuba (*PIW*, 4 March, 1991, p.4), and whose ultimate objective was to reduce the freight costs for both PdVSA and Soyuznefteexport. The arrangement is no longer in force, due to non-fulfilment of its provisos on the part of Russia.
5. This plant also produces 136 tonnes of waxy residue per year.
6. The Colonial pipeline is the setting of the most active gasoline market in the USA. The pipeline is linked with the New York Mercantile Exchange (NYMEX) because all trade in it is hedged in this exchange. On 5 June, 1992, Colonial's link with the exchange became even more direct, because NYMEX launched a gasoline contract based on delivery to the pipeline in Pasadena, Texas. Heating oil is also traded on the Colonial pipeline between October and April. See *WPA*, 11 May, 1992, p.4.
7. The east plant is composed of two atmospheric distillation units, two catalytic reformers, one continuous regeneration reformer, an alkylation unit, one vacuum distillation unit, a gasoil hydrotreater and the petrochemical units. The west plant, owned by General Electric Credit Corporation, and operated by Champlin under a leasing-purchase option arrangement, has a delayed coking unit, a continuous regeneration reformer, a mixed distillate hydrotreater and a desulphurization unit.
8. Typical yields are 60 per cent gasoline, 27 per cent light distillates and 11 per cent fuel oil.
9. Neste Öy later bought out Axel Johnson, and became PdVSA's partner in the venture.
10. The refineries obtain yields of 70 per cent asphalt (general and special purpose varieties), 15 per cent naphthenic lubricants, 12 per cent distillates and 3 per cent of other fuels.

11. The first indications that PdVSA was backing away from a takeover came after Citgo announced its slow down strategy for gasoline sales. The takeover would have led to UnôVen converting to Citgo, and this would have given the company a much greater growth than its stated 8 per cent per annum target (U.S. Oil Week, 21 September, 1992, p.5).
12. The Kuwaitis promptly denied any interest in Uno-Ven.
13. Amoco president William G. Lowrie said after the acquisition that, even though the asphalt operation was profitable, 'the capital needs faced by the refining sector in the US are forcing companies to invest in more strategically key areas' (*PON*, 22 June, 1992, p. 1). Reading between the lines, one can imagine what would have happened with the refinery had Citgo not stepped forth as a buyer.
14. Similar problems had led Exxon to shut down its 400,000 b/d Aruba refinery in March 1985. Exxon tried to get the Venezuelan government to accept a scheme whereby preferentially priced Venezuelan crude would have been used to feed the plant. According to Exxon, this scheme would have preserved a guaranteed export outlet for PdVSA, and saved the Aruban economy from the impact of a closure. The Venezuelan government, however, rejected the proposal.
15. Since 1988, the leasing fee has been $15 million.
16. The current lease will finish in 1994. Barring some major unforeseen developments in Venezuela or the world oil market, it is almost certain to be renewed.
17. In July 1990, PdVSA acquired the assets of the Bahamas Oil Refining Corporation (Borco), a subsidiary of Chevron, for the sum of $120 million. Included in this operation was a 500,000 b/d refinery. However, PdVSA has only used the Bahamas installations for storage, and it has no plans to resume operations in the refinery.
18. The petrochemical complex in the refinery will not fall under the aegis of the joint venture.
19. The sales formula offered by the Venezuelans to Lyondell is apparently based on a guaranteed margin *over* the margin which the refinery could achieve by processing Mexican Maya crude.
20. These lines in *Platt's Oilgram News* are typical: 'By finding a guaranteed home for a percentage of their production . . . [OPEC countries] can short circuit the vicious circle of discounting price to seek more market share for their crude, which in an oversupplied market robs share from somebody else, which in turn leads him to leapfrog the first discounter' (22 June, 1988, p.1).
21. In the case of an integrated producer, this would refer to transfer prices. Lowering one's transfer prices, moreover, amounts to the same thing as price discounting, with the sole difference that it is to oneself that the discount is made.
22. A similar pattern prevails in its European refineries. Most of PdVSA's supplies to Ruhr Öl are constituted by 'medium' crudes (i.e. around 20–25 °API).
23. As Table 8.2 shows, PdVSA's earnings from its export refineries are larger than those of its foreign ventures. However, the margins shown

for the Venezuelan export refineries do not take capital costs into account.

24. Engineering firms, which design and construct grassroots refineries, would probably be less than thrilled to be paid in crude for their services.
25. As Santayana once said, 'those that do not remember the past are condemned to relive it.' The Venezuelan government's lack of money, however, has resulted in some curious contradictions. When in September 1992 the issue of Citgo's sale was revived, Carlos Andrés Pérez said that the sale would also make sense because it would dispel the spectre of possible US government discontent over such a large company being owned by a foreign state firm. This explanation, however, was shot to pieces by the Venezuelan congress (*PON*, 2 September 1992, p.1).
26. Kuwaiti sources denied subsequently that the offer ever took place (*EC*, 26 June, 1992).
27. Citgo accounts for 5.7 per cent of the US gasoline and lubricants markets.
28. PdVSA's internationalization drive has previously come under fire from Venezuelan politicians. For instance, when Carlos Andrés Pérez was making his second (eventually successful) bid for the Venezuelan presidency, he vowed that he would stop PdVSA's overseas acquisition programme dead in its tracks (*O&GJ*, 12 December, 1988). Pérez's campaign promises were the main reason why Steuart Petroleum, a US distributor of refined products, pulled out of a deal that would have given PdVSA a 50 per cent share in the company.
29. And the purchase of the Amoco refinery went ahead only ten days after.
30. Of course, the revenues which these crude volumes would have generated are foregone by the government. However, since the conditions in the acquisition contracts spread out payment over a number of years, the government has generally not felt that, revenue-wise, it is making too many sacrifices.
31. PdVSA expressed an interest in this refinery for quite some time before it actually entered negotiations for a joint venture deal involving the refinery, even though its ownership structure (ARCO owns 40 per cent of the shares, and the rest is traded in the open market) made it more difficult to engineer an acquisition. In July 1991, *PIW* even reported that PdVSA had concluded an agreement with ARCO to lease the latter's share in the refinery (1 July, 1991, pp. 1–2).
32. At the moment, the average gravity of Citgo's feedstock oscillates between 21° and 23° API. Only the Paulsboro refinery's supplies are consistently below 20° API.
33. The very expensive Schwedt upgrading programme would be a prime target for cancellation, if PdVSA were to be asked to pay for its share with cash, instead of crude supplies. Nynäs' expansion, though, seems to be on track. The company has formed a joint venture with Estonian and Russian interests to build a new oil terminal and bitumen storage facility at Tallin, Estonia. Nynäs will hold a 40 per cent share in this new enterprise, called Nybit. Furthermore, if the Dundee and Eastham refinery acquisitions go ahead as planned, Nynäs will eventually hold 17 per cent of the European asphalt market (*O&GJ* 3 August 1992, p.24).

As for refining ventures in the USA, the Lyondell acquisition seems to have closed the chapter on PdVSA's expansion in this market, at least until a new government is inaugurated in the country. The *Bloomberg Oil Buyer's Guide* reported that PdVSA's strategic planning co-ordinator had said that the company would suspend overseas joint ventures for the 1993–8 period (9 November, 1992, p.1).

34. PdVSA does not base this statement purely on the good will of the present and future US governments, however. It has taken good care to protect its investments against possible antitrust problems by means of a dense mesh of legal mechanisms. For instance, when it acquired the whole of Citgo in 1990, the company's new owner – on record – was Propernyn BV, described as 'an indirect, wholly owned subsidiary of PdVSA' (*O&GJ*, 14 January, 1991, p. 37).
35. This will not happen in the near future, however. According to the *Latin American Weekly Report (LAWR)*, opinion polls published on 4 August, 1992 in Caracas, showed that 38.5 per cent of the interviewees would have voted for Lieutenant General Hugo Charez (leader of the February coup attempt) if elections had been held at that point. This is a matter of concern because, up to that point, the maximum popularity level in the polls had been achieved by ex-president Rafael Caldera, with 28.5 per cent (*LAWR*, 20 August 1992, p.1). The coup attempt by the air force, which rocked Venezuela at the end of 1992, is also indicative of how poor the prospects are for any improvement in the delicate domestic political scene.

9 OIL AND THE VENEZUELAN ECONOMY

9.1 Oil and the Illusion of Development

The destinies of Venezuela and petroleum have been inextricably linked in a sometimes uneasy marriage for the best part of the twentieth century. Revenues from oil exports made their first appearance in Venezuelan national accounts in 1917, when the country exported an average of 156 b/d. In that year, the Venezuelan treasury received about 64 million bolívares in fiscal revenues. Oil exports accounted for 0.3 per cent of this total (CEPET, 1989, v. II, p. 493). From this date onwards, the role which oil revenues played in the Venezuelan economy gradually became pivotal: as the Venezuelan export flows expanded exponentially, the bolívar gained value against other currencies. This, in turn, caused the prices of Venezuela's traditional exports (coffee and cacao beans) to rise to a level where they were no longer competitive. Thus, as traditional revenue-earning activities stagnated, and oil-related activities flourished, Venezuela's dependence on oil as a source of government revenue and foreign exchange became nearly absolute. By 1926, oil had displaced coffee as the country's most valuable export commodity and biggest revenue generator; by 1929, it was providing 76 per cent of the country's export earnings and half the government's revenues (di Filippo, 1981, p. 164).

The discovery of oil in Venezuela gave the country an acute case of an economic malady commonly referred to as the 'Dutch disease'. This term refers to the adverse effects that the traded sectors in an economy[1] suffer as a result of the appreciation in the real exchange rate that can follow either a favourable (once-for-all) shift in the production function of a commodity within that economy, or a windfall discovery of new resources; or finally, an exogenous rise, relative to the price of imports, in the price of a particular commodity that this economy exports (Corden, 1984, p. 360). Provided that the income generated by the export shock is spent domestically, such a reallocation hits a country's economy in two ways. First, with a resource effect: 'the marginal product of labour [in the booming sector] rises . . . so that at a constant wage in terms of tradables, the demand for labour in this sector rises and this induces a movement out of [other sectors and into this sector]' (Corden, ibid.) Secondly, there is a spending effect: 'the higher income resulting from the boom leads to extra spending on services [and other non-tradables, raising] . . . their price' while simultaneously increasing the volume of imports (Corden and Neary, 1982, p. 827). The Dutch disease, in other words, mainly affects resource allocation in an economy where progressing and declining sectors co-exist (ibid., p. 825). How distorted this allocation turns out to be will depend, naturally, on both the magnitude of the economic windfall generated by the boom sector, as well as on the size and degree of diversification of the affected country's economy (and also on whatever countervailing policy measures the government of the country takes). When one considers these two aspects, it is not difficult to see why Venezuela's dependence on oil has proved so long lasting (see Figures 9.1, 9.2 and 9.3). First of all, as Noreng rightly points out, 'oil revenues are qualitatively different from other forms of income [because] they represent a rent'[2] (Noreng, 1980, p. 195). Precisely because of this extremely rent-intensive nature, oil is especially prone to cause bad cases of the Dutch disease, more so than other minerals. As Noreng explains:

> For governments, oil revenues represent easy money. Thus, they can use oil revenues to create a comfortable position for themselves. The problem is, however, that within a complex industrial economy, the ability to absorb a sudden influx of easy money is limited, so that oil revenues tend to become a substitute for other income rather than a supplement. Consequently, the net short-term gain may be less than large oil revenues indicate in a dynamic perspective; the short-term use of rentier income may compromise the long-term generation of other forms of income [given the wilting of other sectors of the economy] (ibid.).[3]

It is important to remember that the most famous cases of Dutch disease have been associated with reversal in the terms of trade of small- or medium-sized industrial economies.[4] Venezuela, however, acquired an ante-industrial form of the disease, which made it all the more pernicious. Indeed, one can say that Venezuela's Dutch disease in the early 1920s, which conferred great strength on the bolívar, acted as a form of industrial contraceptive device (which prevented non-oil related industries from taking root in the country), and stunted the growth possibilities of its agricultural sector.

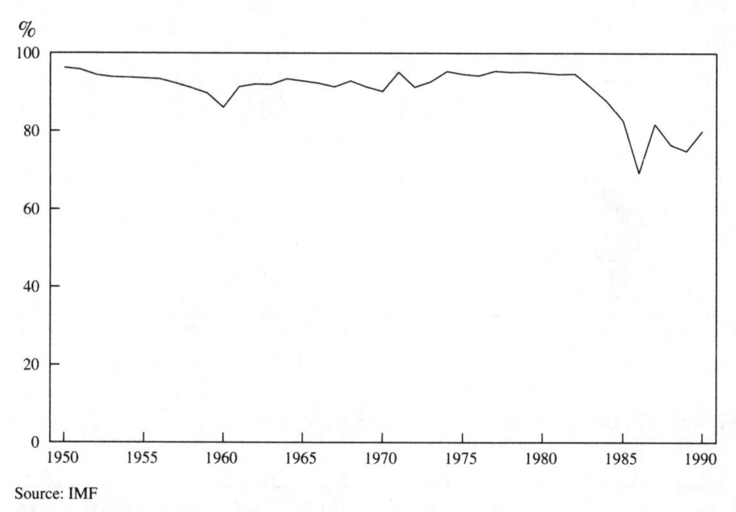

Figure 9.1: Oil Exports as Per Cent of Total Exports.

For a long period of time, this state of affairs was not seen as unduly troublesome: the Venezuelan government simply used a part of its rent-derived purchasing power to buy foreign food, manufactured goods and services. However, some people insisted that using oil revenues to diversify the economy was a task that could not be put off. For instance, Arturo Uslar Pietri warned that Venezuela, if it did not mend its ways, would become 'a passive and unproductive country, a huge parasite, totally dependent on the oil industry, whose affluence would prove to be temporary and lethal, thus making a dreadful catastrophe inevitable sooner or later' (Werz, 1990, p. 189). Uslar Pietri emphasized 'the importance of investing the wealth gained by the destructive activity of mining in building up a productive agricultural industry' (ibid.). In so doing, he set the agenda for all the civilian governments to come, and produced an image of the oil industry which, as we shall see later on, still holds sway in the minds of most Venezuelans.

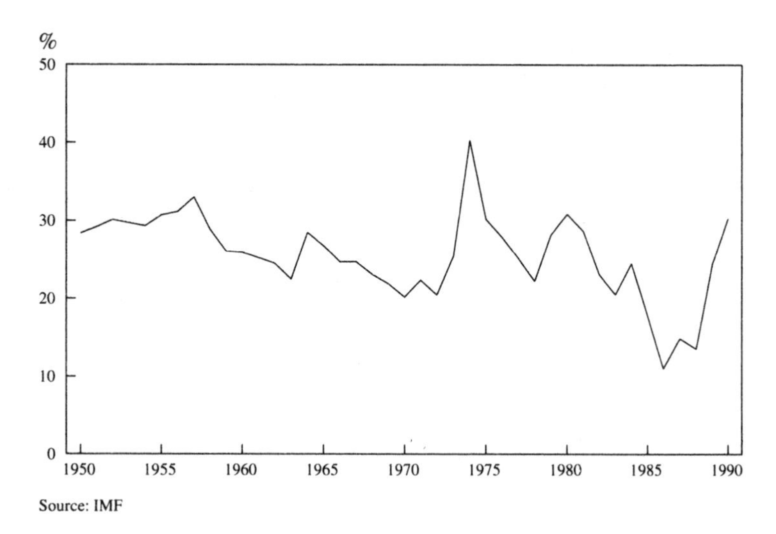

Figure 9.2: Oil Exports as Per Cent of GDP.

After 1958, the new civilian leaders finally decided that the international division of labour which relegated them to the status of raw material exporters was untenable, and that the time had come to make fundamental changes in their methods of capital accumulation and trade management. To achieve this, not surprisingly, they chose to pursue an import substitution strategy, to be conducted primarily by means of barriers to trade.[5] A high tariff wall, they reasoned, would protect the nascent Venezuelan industry, during its start-up phase, both from competition from abroad and from the strength of its own currency. And oil revenues would take care of the rest.[6]

The Venezuelan leaders' expectations concerning the potential of oil to serve as an instrument for overcoming underdevelopment were linked to a conception which saw the state as the centrepiece of the process of capitalist development (Werz, 1990, p. 182). Although this conception was the norm

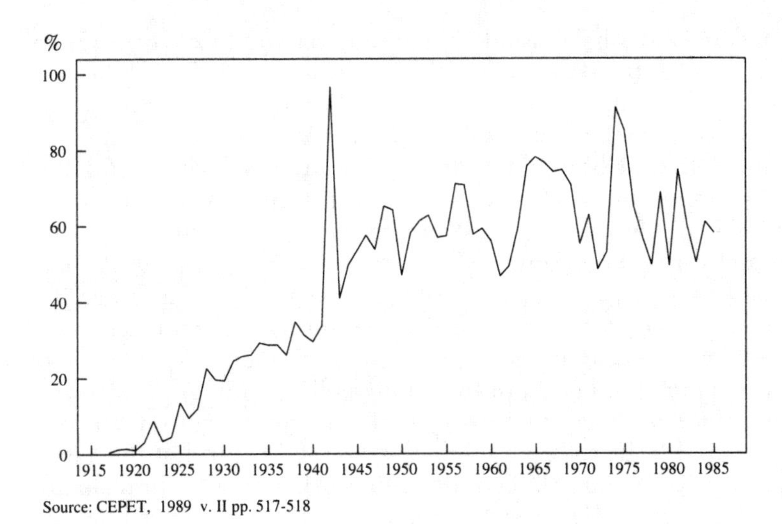

Figure 9.3: Contribution of Oil Export Revenues to Total Government Fiscal Revenue.

– rather than the exception – in the other countries of Latin America at the time, it was embraced with particular fervour in Venezuela, where conservative and social democrat politicians alike praised the development potential of an active state nurtured continuously with a steady flow of oil revenues. A priori, this conception of development is reasonable enough; after all, oil revenues seem like an ideal vehicle for releasing the main constraints on economic growth in underdeveloped economies: a lack of domestic savings, the insufficient availability of foreign exchange and insufficient fiscal revenues. It should come as no surprise then, that during the 1960s and 1970s, all of 'the capital-short and aspiring Third World planners . . . kept telling themselves (and each other) that if only they had . . . black gold, the magical *élan vital* for their economic take-off would be close at hand' (Amuzegar, 1982, p. 814).[7]

Although during the 1960–73 period, Venezuela's *per capita* income was just below that of the least advanced West European countries, the Venezuelan political leaders never made any claims that their country had definitely left underdevelopment behind. Nevertheless, statistics like this one made them fall prey to the illusion that it was on the brink of a major economic breakthrough, which would take Venezuela into the league of developed nations. When the 1973 oil shock occurred, the size of the windfall associated with it (calculated by Gelb and Bourgignon to have been 20 per cent of GDP – excluding the windfall – between 1974 and 1978; Gelb *et. al.*, 1988, p. 295), was seen by many as the element that would finally rid the country of the constraints which, until then, had kept it on the wrong side of the development fence. Of course, there were those who, like former oil minister Pérez Alfonzo, took a far less sanguine view of things. Pérez Alfonzo, in a much publicized tirade, vilified oil by calling it 'the devil's excrement'. According to the man that had struggled for a good part of his life to bring oil under the total control of the Venezuelan state, the oil windfalls would bring nothing but calamities to Venezuela: waste, profligacy, corruption, and above all, debt. Pérez Alfonzo's tirades were not pessimistic just for the sake of it. Indeed, a structuralist revision of the Venezuelan economy told a very different story about the viability of development from the one embraced by the politicians. For instance, in common with the cases of other import-substituting countries, Venezuela's agricultural sector, performance-wise, was lagging far behind the rest of the economy. Also, there were virtually no non-oil exports, because of the anti-export bias implicit in import substitution schemes,[8] and heavy regulation rendered most of the non-oil economy non-traded (in the sense that there was little effective competition with imports). Unemployment stood at 8 per cent of the labour force in 1971,[9] and Venezuela's income distribution was among the worst in the world: according to di Filippo, 50 per cent of the Venezuelan population received 14.3 per cent of national income; the *per capita* income of this half of the Venezuelan population came to about a sixth of the *per capita* income that Venezuela's richer half received (di Filippo, 1981, p. 166).

The Venezuelan presidential race of 1972 saw the most sensitive and visible of these negative economic aspects (poverty and inequality) assuming the centre stage in the political debate. Acción Democrática's candidate, Carlos Andrés Pérez (the eventual winner of the election) stressed the government's obligation to banish these two blights on the country's prosperity. When the oil price quadrupled in 1973–4, Pérez – seizing the chance to make good his promises – started a massive spending programme with three main objectives in mind: boosting the output of the public sector, raising wages and increasing the overall level of employment in the economy. The first of these objectives was the most important part of Pérez's strategy, because the achievement of the other two aspects was, to a degree, contingent on its success. Because of this, the public sector assumed an even more important role than the one it had before the oil shock, and this fact was given legal expression in Venezuela's fourth National Development Plan. According to the guidelines established by the plan, the manufacturing activities of the private sector were to be oriented primarily towards import substitution, while the massive state companies would be responsible for industrial exports. This fact, added to the degree of concentration present in Venezuelan industry, made the Venezuelan non-oil economy overtly dependent on the performance of a few major firms (and since six of the country's largest ten industrial companies were state owned, any financial difficulties which they experienced had a direct – and sometimes massive – impact on the government's budgetary position). Unfortunately, the state-owned manufacturing companies did not live up to their output promises, and as a result of their failure to become financially viable entities, they accumulated a sizeable deficit, and contributed with a substantial proportion of the increase in the Venezuelan external debt (particularly in short-term credits).[10]

Subsidies were a central part of the strategy for the distribution of oil wealth. Transfers to households averaged 7 per cent of the government budget (windfall excluded) over the period 1973–8. These transfers comprised sums set aside to cover the deficit of the Marketing Board for Agricultural Products (which amounted to about 1 per cent of GDP), from 1974 onwards. This deficit arose as a result of a price controls and subsidy policy imposed by the government to protect low income groups from the effects of inflation. A negative aspect of price controls (which eventually encompassed four-fifths of all wage goods), was that they squeezed the profits of the entrepreneurs who produced the controlled goods. This, in turn, caused private sector capital to be diverted to areas other than manufacturing, notably services.

The spending spree which began in 1974 had a short-term multiplier effect. However, as the economy began to approach capacity, it showed definite signs of overheating and capacity strain. By 1977, the growth of imports had led to a trade deficit being recorded (quite an ominous development, especially when one considers that, as a result of the oil price increases, the Venezuelan terms of trade had experienced a dramatic improvement). Inflation accelerated, and non-oil GDP growth rates began to decline noticeably (from the 11 per cent registered in 1975 to 3.7 per cent in 1978). The second oil shock provided the government with an excellent opportunity to redress past mistakes. Sadly, it was an opportunity which Venezuelan politicians, always unwilling to introduce unpopular measures, chose not to take. Good intentions abounded, of course. The Christian Democrat COPEI party (whose candidate, Luis Herrera Campíns reached the presidency in the wake of massive political discredit for Acción Democrática) preached a platform of austerity and fiscal moderation. However, it did not try to go to the roots of the structural problems riddling the Venezuelan economy (low productivity, inefficiency in the public sector, and so on). Therefore, the new president's policies became a matter of pouring in good money after Pérez's bad, in the hope that the awaited industrial 'take-off' would finally occur. This, clearly, was a recipe for disaster. The great public projects inaugurated by Pérez consistently failed to perform (for instance, the annual production of the aluminum and steel state enterprises, which together had swallowed much of the manufacturing investment by the Venezuelan state, amounted only to about 40 per cent of installed capacity): quite unsurprisingly, they incurred massive operating losses which had to be alleviated with direct government transfers. To make matters worse, the ceiling on interest rates, an unrealistic exchange rate and the profit squeezes due to price controls fostered a sensible drop in the private sector's saving rate and massive capital flight; according to Mayobre (1987, p. 142), from 1978 to 1981, investment by the private sector in Venezuela shrank by 50 per cent in real terms. This provoked the government into spending even more, in a last-ditch attempt to fend off recession.

Things came to a head after 1981. A weakening international oil market caused Venezuela's export income from oil to fall by approximately one-fifth. This made it clear that, unless the earnings from oil went back to previous levels (an extremely unlikely eventuality), the burden of balancing the external account and sustaining the country's economic growth would fall on other sectors. As the oil market continued to weaken, however, the Venezuelan economy revealed itself as incapable of coping with the stress. Against an alarming backdrop of private capital flight, the government surplus of 2 per cent of GDP in 1981 became a deficit of 2.6 per cent in 1982. Towards the end of the year, with the situation becoming more desperate, the government imposed import restrictions and raided PdVSA's foreign exchange reserve chest. By 1984, unemployment was at a 12.1 per cent level and the informal sector of the Venezuelan economy was providing occupation for 43.8 per cent of the labour force (Yáñez Betancourt, 1987, p. 729).

During the 1950–73 period, when the average price of oil was lower than $2 per barrel, Venezuela's GDP grew at an annual rate of 6.4 per cent (15 per cent above the growth rate of the rest of Latin America). After the first oil windfall (which spanned the years 1973–80), when the price of oil averaged $4 per barrel, the country grew at a much inferior 4.1 per cent annual rate (when the rest of Latin America grew at a 5.4 per cent annual rate, and Venezuela's population increased at a 3.6 per cent annual rate). The second oil windfall (from 1981 until 1985), which increased the average price of oil to $26 per barrel, coincided with an even greater deterioration in the Venezuelan economy. During this period, the country's GDP declined at a 1.3 per cent annual rate. Also, by 1979, the country's debt, which had been less than $1 billion at the beginning of the 1970s, was only just under $25 billion (Yáñez Betancourt, 1987, p. 742). These figures serve as an indication of the Venezuelan government's economic policy during the windfalls. Gelb and Bourgignon summarize quite well the fiasco of the great prosperity promised by the oil shocks against the dearth of actual results achieved when they say: '[By the time the decline in the price of oil set in, the Venezuelan economy] was one third smaller than it would have been if [its] expansion had continued in line with the pre-1973 trend. . .The overall picture that emerges is one of dramatic failure of economic policy under the most favourable conditions. Nothing appears to have been gained from the windfalls in terms of non-oil GDP during 1973–1982' (Gelb *et. al.*, 1988, pp. 321–2).

Jaime Lusinchi, successor to Herrera Campíns, did not live up to his electoral mandate of putting the country back on the right track with a strong dose of fiscal discipline and austerity medicine. On the contrary, his administration maintained the subsidies for many foodstuffs, medicines, and similar basic goods. The large fiscal and balance of payments deficits which resulted from this policy in the process, nearly exhausted Venezuela's once abundant international reserves (Ochoa, 1992, p. 23).[11] When the stringent economic adjustment measures recommended by the IMF (increasing domestic fuel prices and public transportation fares by about 100 per cent) were finally implemented in 1989 (at a moment when the annual rate of inflation had reached an all-time high of 89 per cent), Venezuela erupted in a series of nationwide riots and looting from 27 February to 3 March, 1989, which left scores of people dead and thousands injured. The riots expressed the pent-up frustration of a people who could not understand how it was possible that after 'the largest non-violent transfer of wealth in human history' (Schneider, 1983, p. 1), their nation – a prime recipient of this wealth – had emerged as underdeveloped and dependent as it had been before the oil shocks. The 1989 civil unrest, in other words, was a searing indictment against 'establishment' politicians from AD and COPEI, who had turned the citizens of what was thought to be a reasonably prosperous country into the biggest debtors in Latin America.[12] As for the success of the Venezuelan industrialization drive, intended to wean the country from oil, facts speak poignantly for themselves: in 1991, PdVSA was still responsible for 24 per cent of the country's GNP, 86 per cent of its foreign exchange earnings and 83 per cent of the government's tax revenue.

9.2 Revenue Needs, Taxes and PdVSA

The Venezuelan government, as we have seen, was not successful in its task of using oil revenue to propel the country's economy along the path to development during the period of the oil windfalls. When the decline in the price of oil set in during the middle 1980s, and the government's financial situation became untenable, it enacted policies which not only did nothing to alleviate Venezuela's economic predicament, but have also come to threaten the viability of the Venezuelan oil industry itself.

In November 1991, as the Christmas recess of the Venezuelan congress approached, Hernán Anzola, a member of a congressional committee on energy (and former oil minister), accused the political establishment of gross irresponsibility in the management of the oil sector. Anzola pointed out that 'neither the government nor Congress [had] made any effort to introduce the reforms that are needed to eliminate the discriminatory tax treatment of the oil industry' and that the government had made 'no effort to introduce a progressive reduction of export reference values' (*PON*, 20 November, 1991, p. 5).[13] 'The government's voracious appetite for revenue', Anzola said, '[was] threatening the future health of the nation's oil industry' (ibid.).

When viewed in the context of PdVSA's 1991 performance, Anzola's words seem excessively alarmist. After all, during this year, PdVSA became the third largest oil company in the world, relegating the Exxon Corporation to fourth place (*OB*, April, 1992, p. 52).[14] It increased crude production by 240,000 b/d (production reached a peak of 2.5 mb/d, its highest level in 16 years), refinery throughputs by 97,000 b/d, crude exports by 237,000 b/d and proved oil reserves by 2.7 billion barrels. However, underneath these favourable numbers lies the indisputable fact that PdVSA is a giant with clay feet, whose future plans to develop a much larger presence in the global energy market[15] depend, first and foremost, on whether it is able to consolidate its financial base in order to raise the enormous sums of money that its diversification efforts will demand. But this consolidation can only come about if the government changes the way it levies taxes on PdVSA's activities.

The problem with PdVSA's tax burden, simply put, is its magnitude. The company, being a 'legal entity engaged in the exploitation of hydrocarbons and similar activities, or in receipt of royalties and similar income derived from hydrocarbons and related activities' is subject to a 67.7 per cent income tax rate.[16] According to the government, this tax burden is not excessive, because the company's earnings are such that it can easily manage to finance investment out of its cash flow. However, this argument omits a very important point: the government's calculations of the income tax payable by PdVSA are based on a 'fiscal value' for crude exports (not actual realized prices), which is equal to the invoice value of the petroleum multiplied by a factor fixed by the government (see Figure 9.4). By means of this simple operation, the Venezuelan government increases PdVSA's tax burden by about 13 per cent.[17]

One would expect this kind of taxation rate to leave the company with extremely reduced funds for investment. A look at PdVSA's annual reports confirms this suspicion. During the last few years, for instance, the company's net working capital

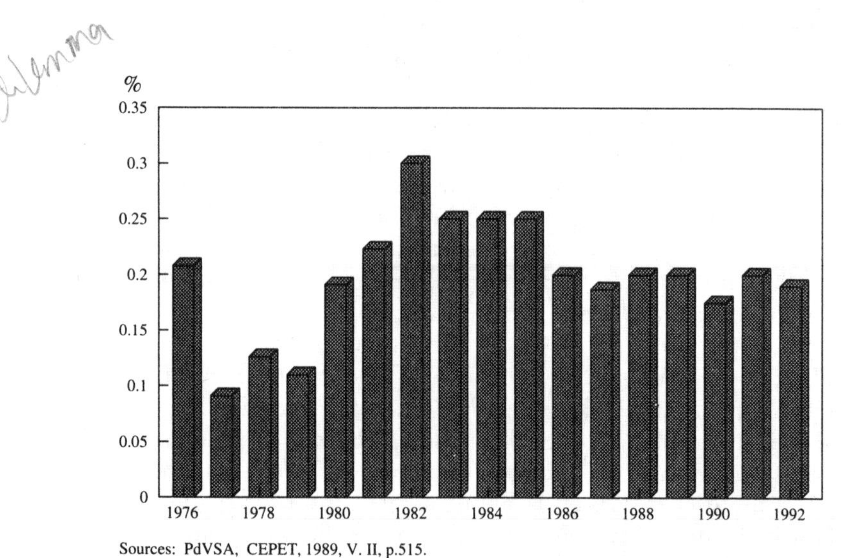

Sources: PdVSA, CEPET, 1989, V. II, p.515.

Figure 9.4: Tax Reference Value of Oil (Percentage Added to Realized Invoice Value).

(current assets plus negotiable investments less current liabilities) has never gone beyond the equivalent of a few weeks' sales, and on occasions it has even been negative. This has left borrowing as the only viable alternative for PdVSA to finance its needs.[18] PdVSA's repeated requests to the government for a lessening of its tax load and the abolition of the export reference price, then, come as no surprise. Unfortunately, the government's refusal to grant PdVSA's requests is just as understandable.

According to Adelman, oil-producing countries resemble leveraged corporations (firms with large prior claims on their gross income due to high levels of debt or high rentals) with the difference that 'governments tend, even more than private persons, to anticipate and overspend future income, saddling themselves with debts or other commitments. The greater the oil income, the more they leverage it' (Adelman, 1986, p. 22). This simile fits Venezuela perfectly: today, the government's oil revenues are not nearly sufficient to meet its spending needs, the servicing of its foreign debt and the investment needed to promote economic growth in the country.[19] Hence its refusal to consider implementing new tax schemes, whose first effect would be to diminish the level of its income.[20]

The abortive military coup of February 1992 has succeeded in making PdVSA's financial position even more uncomfortable than before, severely jeopardizing its chances of becoming a global energy conglomerate by the end of this century. The relatively tepid popular reaction against the coup confirmed the worst fears of the government; namely, that its legitimacy, as well as that of the country's democratic institutions, has been eroded to an unprecedented extent.[21] The 'Bolivarian' demagoguery of the rebels struck a chord of sympathy in a people who have seen their purchasing power being relentlessly undermined by inflation, the quality of public services diminishing steadily, and widespread corruption in the central and local governments going unchecked. Thus, the coup made it clear to Venezuelan leaders that the dyke of popular discontent was dangerously near its bursting point, and that, in consequence, social and welfare expenditure would have to be increased, if only as a stop-gap remedy to this general social malaise. Unfortunately, since the government has to remain within the IMF-mandated limits on public expenditures and borrowing, the only way in which it could raise enough money for these expanded social expenditures was by siphoning it from something else. That something has turned out to be PdVSA's investment budget:[22] the government has ordered the company to scale back its non-crude operations (such as LNG, Orimulsion, coal and petrochemicals), abandon its plans to boost its crude output capacity, and find foreign partners for its proposed refining ventures in Venezuela. Although it is still too early to tell what the full effect of these budget cuts will be on both the upstream and downstream sectors of the oil industry, it is probably fair to say that, in the future, nobody will be able to take the health of the Venezuelan energy sector for granted.[23]

9.3 Oil: A New Platform for Industrial Development?

As this chapter has shown, the government's attempts to 'sow Venezuela's oil', and turn the country into a medium-size industrial power, have not been very successful. This failure can be traced to the conception of the oil industry which has dominated Venezuelan oil policy since Uslar Pietri coined the phrase 'to sow the oil', and which has always seen the industry's *raison d'être* as the extraction of mineral rents, which have to be channelled by the government to other sectors of the economy in order to lead the country to an economic state where activities using reproducible factors of production – and not based on natural resource exploitation – are the main sources of income. However, the process whereby 'a portion of the income stream earned in the enclave is siphoned off [by a country's government] for the purpose of irrigating other parts of the economy'(Hirschman, 1981, p. 69) has a crucial shortcoming; namely, that there are no signals for the governmental agents to see which tasks should be conceded maximum priority and this makes 'the possibility of either faulty investment, or a great deal of leakage on the way' a very high one, and even more so when it is compounded by other factors which diminish the effectiveness of public spending, such as rent seeking activities, or the inadequate and necessarily limited planning capacity of the public sector (1981, p. 69). As we have seen, Hirschman's conclusions, rather unfortunately, fit the case of Venezuela during the oil windfall period perfectly.

Although the point of view which considers that rent from oil should find its way to 'productive' sectors of the economy by means of fiscal policy still predominates in Venezuelan policy-making bodies, the last few years have witnessed a gradual decline in its popularity and an increasing awareness of the fact that it seems to represent an empirical dead-end. Because of this, there have been proposals to substitute this apparently sterile approach with a new one, which sees the oil industry as a nucleus of industrial activity in its own right. The new approach considers that oil should function as the focal point of a medium-term industrial development strategy, whose central objective would be to concentrate policy incentives and support infrastructure on industrial areas where Venezuela enjoys either an initial comparative advantage or at least has a realistic chance to develop one. The developmental emphasis, in other words, would move towards developing upstream and downstream industries connected to the basic activity of petroleum production. The proponents of this view contend that, upstream, PdVSA's demanding specifications have turned the Venezuelan companies that supply it into entities able to compete in the international market. Downstream, production linkages would be established in the petrochemical and chemical industries, two areas where Venezuela enjoys sizeable potential comparative advantages because of the availability of comparatively cheap feedstocks and qualified human resources.

This 'new view' of the oil industry serves its interests better, if only because it explicitly recognizes that it makes no sense for PdVSA to be bled dry. Whether it will eventually hold sway in Venezuela is, at the moment, anybody's guess. After all, in circumstances of extreme financial duress, even a government that considered this view as being sensible might ditch it as unwanted ballast. In the meantime, while PdVSA waits and hopes for the best, it will have to find ways of getting around the cash starvation that besets it. General Alfonzo Ravard once said, '[financial] self sufficiency is indispensable for our freedom of judgement. If industry [does] not have it, it would have to ask the state for funds or borrow abroad. [The] first alternative [is] contradictory, because the industry supplies most of the funds to the state, [while] the second alternative [leads] to a serious erosion of our freedom of decision' (Coronel, 1983, p. 227).

Notes

1. Traded sectors are those that produce goods whose domestic prices relate to international prices, as a result of the possibility of goods going across international borders. Non-traded sectors, in contrast, serve markets that have to be cleared by domestic price movements exclusively.
2. The term rent is used here in a Ricardian sense, i.e. as the difference between the market price of a commodity and its production costs, with capital and labour being paid at market rates. The Ricardian rent element in the price of low-cost oil – which has accounted for the majority of all oil produced – has always been quite notable. Since 1973, however, it has grown to massive proportions, far larger than those recorded for any other product.
3. Noreng ingeniously described the effect of oil discoveries on a country's economy as a process akin to the one described by Gresham's Law: 'Bad money drives out good'. Expressed in a suitably aphoristic fashion, Noreng's maxims are: 'Easy money drives out difficult money; rentier income drives out productive income; petroleum revenues drive out industrial income' (Noreng, 1980, p. 195).
4. Like Norway or the Netherlands. Indeed, the term Dutch disease was coined to describe the effect which the discovery of the Groningen gas fields had on the economy of the Netherlands (Corden, 1984, p. 359).
5. Venezuela's substitution of imports began during the second world war, when shortages caused by the conflict prompted the growth of domestic industry (the same thing happened in other countries, such as Mexico). However, a coherent import substitution *policy*, using tariffs and trade barriers, was only adopted later, after the end of the war.
6. Tugwell (1975, p. 167) called this approach the 'fiscal saturation' policy.
7. Developments in Venezuela during the 1960s and 1970s seemed to confirm these conclusions: during the period 1960–69, Venezuela's non-oil GDP grew at an average annual rate of 6.5 per cent (Hausmann(b), 1990, p. 3), while manufacturing grew at a 7.1 per cent annual rate from 1957 to 1973 (Echevarría, 1985, p. 29).
8. For a lucid explanation of how import substitution policies work, as well as how they bring about these undesirable consequences, see Little, Scitovsky and Scott, 1970, *passim*.
9. This figure, however, does not accurately reflect the real situation of the Venezuelan working classes at the time, since it made no provision for underemployment (which was widespread).
10. Venezuela's Credit Law of 1976 required congressional approval for all public sector borrowing, except when the funds involved were to be

used as short-term working capital. This type of borrowing only needed approval from the finance ministry, which in those days was quite happy to oblige the companies' requests. Thus, public companies borrowed abroad to cover their operative losses, in the process contributing to a massive increase in the short-term debt. A few years later, the structure of the gross debt portfolio would give rise to grave problems, because Venezuela would find herself extremely hard pressed to come up with enough cash to meet its short-term financial obligations (Mayobre, 1987, p. 144).

11. The Venezuelan Central Bank abandoned the traditional unified and fixed exchange rate system in favour of multiple rates only after the loss of more than $10 billion in international reserves (Hausmann(a), 1990, p. 6).
12. With a debt of $34 billion at the end of the eighties, Venezuela had the biggest *per capita* external debt of any Latin American country. Debt totalled 54 per cent of GDP (151 per cent more than the value of total Venezuelan exports), and debt servicing ate up 55 per cent of the country's export earnings (Mayobre, 1987, p. 140).
13. See below for an explanation of the export reference price.
14. In terms of operating indicators like total liquids reserves, and total refining capacity, but certainly *not* of financial indicators like revenues, total sales, or profits.
15. According to the spirit of PdVSA's 1991–6 plans, by the year 2000, the company was to have become an energy conglomerate, deriving more than 25 per cent of its income from non-oil operations, including natural gas, petrochemicals, Orimulsion and coal.
16. Its 'nonhydrocarbon-source income (excluding dividends)' is also subject to the same taxation rate.
17. 'The National Executive is authorized to establish tax values of hydrocarbons up to 20% of realized values.' Before 1981, there was no defined percentage by which the tax values could exceed actual values. After that date, the Venezuelan income tax law was amended to include such limits. From January to August 1990, these percentages were fixed at 15 per cent of the average sales realization price per barrel. After August 1990, they were fixed at 20 per cent of realized value (PdVSA, 1990). As a result of the financial squeeze affecting PdVSA in early 1992, the export reference values were lowered to 19 per cent in June and 18 per cent from October onwards.
18. In 1988, two US banks granted PdVSA a loan based on future oil sales through its Lake Charles and Corpus Christi refineries. The deal, according to Evans, 'essentially amounted to immediate availability on cash in return for accounts receivable on the sale of about 300,000 to 325,000 bpd of crude up to 1995' (1991, p. 392). This operation contradicted PdVSA's stated policy not to use oil to repay loans.
19. Total investment in 1991 was supposed to be 15 per cent of GNP. The private sector's share of this investment, however, will account for, at best, only 7.5 per cent of GNP.

    The Venezuelan exchequer's plight is not only a function of uncontrolled spending by the government. Falling oil prices and population growth are also to blame. This can be illustrated with a simple exercise. In 1974, for instance, the government's fiscal earnings from oil totalled $9.3 billion. Projecting these numbers forward into 1991 (taking into account a cumulative inflation rate in dollar terms of around 150 per cent) results in a figure of $23.3 billion. Had it received this amount of money in 1991, the government would have been at the same earnings level as it was in 1974. However, this operation does not take into account the 65 per cent growth in Venezuelan population since 1974. When population is factored in, the result is that, if the government had wanted to earn as much oil revenue *per capita* in 1991 as it did in 1974, its foreign exchange earnings from oil exports would have had to be around $38.4 billion (*PON*, 15 November, 1991, p.5). Unfortunately, however, PdVSA's 1991 income was only $14 billion (*OB*, April 1992, p.52).

20. In mid-1992, the government began studying a new tax mechanism for the oil industry akin to the one used by Canada and the UK. The mechanism would combine a fixed tax ceiling with a variable royalty (*PON*, 29 May, 1992, p. 6). To alleviate PdVSA's short-term liquidity problems, in June of 1992, the government consented to give PdVSA a tax break by lowering the export reference values to 19 per cent in June and 18 per cent from October onwards. Furthermore, according to PdVSA President Gustavo Roosen, the Venezuelan executive has accepted a draft bill prepared by PdVSA that contemplates the gradual phasing out of the export reference values over the 1992–5 period (*PON*, 23 October, 1992, p.4)
21. There have even been rumours that Carlos Andrés Pérez might step down from the presidency before his term in office is over.
22. Even before the coup, Hernán Anzola had predicted that the onset of a new electoral cycle – municipal and gubernatorial elections are scheduled for 1992, presidential elections for 1993 – would be likely to accentuate the government's need to tax the oil industry to the utmost (and hence increase the damage to it, because 'populist pressures and electoral manipulations' would push the country 'into a series of heavy social expenditures for which there are no financial resources available' (*PON*, 20 November, 1991, p.5). Unfortunately, Anzola's words have proved quite prophetic.
23. Especially since the expansion of crucial areas like LNG, Orimulsion, coal and petrochemicals will come to depend mainly on foreign investment. However, even if it retreats from what the government considers non-core businesses, PdVSA will still have a rough time ahead, finance-wise. According to the company's current president, Gustavo Roosen, PdVSA must invest at least $2.2 billion per year just to maintain crude output potential at its 1991 level (*PON*, 8 June, 1992, p.5).

10 THE FUTURE OF VENEZUELAN OIL

In the year 2014, the history of oil in Venezuela will reach an important milestone; namely, the one hundredth anniversary of the discovery of commercial deposits in the county. Furthermore, the fortieth anniversary of the nationalization of the Venezuelan oil industry will take place only two years later. Unfortunately, however, the current situation of the oil industry is such that there is a real chance that by then the country will be in the surprising and uncomfortable position of being a marginal crude oil producer (as PdVSA's president Gustavo Roosen has warned on occasion). It is not difficult to understand why Venezuela's status as a first rank member of the league of major oil producers may be jeopardized early next century. The reason is quite familiar – dwindling reserves of conventional light and medium crudes.

As will doubtlessly be remembered, the very heavy nature of Venezuelan incremental oil production was the main problem confronting the industry in the years after nationalization. However, the mid-1980s discovery of light and medium crude oil reserves in the eastern and western parts of the country put these troubles to rest for a while. The crucial importance of these finds for the development of the Venezuelan oil industry cannot be emphasized enough. Thanks to the output from fields like Ceuta, El Furrial and Guafita, Venezuela was able to arrest the inexorable decline in its light and medium crude oil production and, although production from these fields has never accounted for much more than 10 per cent of total production, their output postponed PdVSA's hour of reckoning with the predominantly heavy, metal laden and sulphurous Venezuelan crude base by at least 10 years. This simplified PdVSA's existence no end, because in 1985 it was in a very poor position to produce and process rapidly increasing volumes of very heavy crude.[1] As it turned out, PdVSA took full advantage of this reprieve by boosting its light and medium crude output (thus augmenting its revenue and reducing its costs), while gaining valuable experience in many areas of heavy oil exploitation, such as enhanced oil recovery, artificial lift, drilling, well completions and workovers, and the handling, treatment and transportation of extra heavy crudes.[2] This expertise has already yielded substantial and tangible results, for example the development of Orimulsion.

Unfortunately, the reprieve afforded Venezuela by the discovery of these fields was only temporary (and relatively short to boot): production of the 1980s vintage fields is expected to peak shortly after the year 2000, and to start declining steadily afterwards. Thus, towards the end of the century, Venezuela finds itself in a situation similar to the one it had to tackle in 1985. Its declining output of light and medium crude oil means that Venezuela has either to find new reserves of conventional light and medium crudes, or to turn its attention once again towards the vast potential of the Orinoco Oil Belt.

On the surface, it might seem that the current situation of the Venezuelan oil sector is not unfavourable to such efforts. After all, the average API gravity of current crude production is appreciably higher than it was in the early 1980s. Also, the Venezuelan refining system today is 180 degrees removed from what it was in 1976; its erstwhile simple configuration has given way to a high conversion, complex configuration. Finally, on a corporate level, one need only compare the confusion of companies large and small which constituted PdVSA in its beginnings, with the sleek energy conglomerate it is now in order to get a real measure of the Venezuelan oil sector's progress.[3] For all these reasons, one might be excused for thinking that the Jeremiahs who predict the precipitous decline of the Venezuelan oil industry are adopting an overtly alarmist position. In fact, however, the reverse seems to be true.

According to a recent study commissioned by PdVSA and undertaken on its behalf by BP,[4] current reserves in existing production areas can sustain Venezuela's crude production until the year 2000. After this date, the study predicts that output from these will decline quite rapidly (from 2.4 mb/d in 1992 to barely 800,000 b/d by 2030). I broadly agree with these figures, and also with the conclusion that the study derives from them: 'Venezuela must find enough oil in the next 20–30 years to replace over 2 million b/d of depleted capacity' (*PON*, 28 January, 1993, p.4). In order to meet this quite daunting goal, as has been said, Venezuela can choose between devoting the lion's share of its attention towards developing the extra-heavy oil deposits of the Orinoco (the first option discussed below) or concentrating upon finding new light and medium crude reserves through strategic association schemes with foreign oil companies or contractual partnerships (the second and third options).

The first option is mainly advocated by those who feel that Venezuela's conventional reserve base is nearly exhausted, but it is a difficult enough enterprise to discourage the major oil companies of the world, on various grounds. First of all, there is the fact that environmental regulations all over the world will become more restrictive in the years to come. This green trend may complicate or even jeopardize the future of the most straightforward way of exploiting the Orinoco Belt yet devised (the production of Orimulsion). This also makes it virtually certain that, in years to come, PdVSA will have to invest heavily in costly desulphurization technology for Orimulsion, since this will be the only way for the product to continue making inroads in the fuel markets of industrial nations. Opening up the Orinoco also poses substantial technological problems, because in order to transform the oil belt into a genuine oil province, PdVSA will have to plough a lot of resources into improving the very expensive – and, as yet, basically unproven – hydrogenation technology for extra-heavy crude upgrading (since this technology is less expensive than the traditional carbon rejection processes in terms of the proportion of the crude that ends up as an undesirable by-product like coke). However, even though – technologically speaking – PdVSA is better positioned than most of its peers in the third world to cope with these challenges, the obstacle of coming up with the money needed for the Orinoco mega-projects[5] seems to have an air of insurmountability about it. This is particularly true given the apparent consensus among the major oil companies of the world that returns commensurate to the extremely high start-up costs of the projects would only stand a chance of being achieved starting at a price of $25 per barrel of oil at constant 1992 prices; this is about $10 per barrel more than that received by Venezuela in 1992 for its heavy crude exports. Thus, on the strength of the evidence presented above, it seems logical to postpone for a while any plans of turning the Orinoco Oil Belt into the premier oil patch of the world, and focus instead on exploration efforts, and Orimulsion development. This last assertion, of course, would not be valid if Venezuela is found to be 'running out of oil' as some people contend – if this should occur then development of the Orinoco would be the only viable alternative for the country. However, as Adelman has repeatedly said, the quantity of any mineral remaining in the earth's crust is a completely irrelevant fact when taken on its own. Simply put, if the cost of inventory replenishment and extraction of new resources should exceed the market price for the commodity, investment would cease, and this would transform whatever mineral remains in the deposit into a geological curiosity; reserves, in other words, represent an inventory that is replenished through investment. It is a little premature to talk about the imminent demise of an oilfield or oil basin merely on the grounds that its reserves are being depleted at a very fast rate, even if it has been under exploitation for a long time. Venezuela's prospects as a conventional oil province, therefore, should not be written off too quickly. As Adelman wrote in 1989:

> The USA is the extreme example or paradigm of the very old oil province. No area in the world is as drilled-up today as this country was (excluding Alaska) in 1945. 'Remaining recoverable reserves' were 20 billion barrels. In the next 42 years, the 'lower 48' produced not 20 but 100 billion, and had 20 billion left (1989, p.19).[6]

In sum, one can say that there are several reasons why the policy choice of an all-out, no-holds-barred, development of the Orinoco is an ill-advised course of action for Venezuela, even under the assumption that the country could afford such development. Thus, when considered within the context of PdVSA's current financial outlook and the apparent unwillingness of foreign oil companies to co-operate in the process, the case against Orinoco development seems to be overwhelming. None the less, one should stress that PdVSA's continued insistence on opening up the Orinoco to commercial exploitation is quite logical – even though it contradicts BP's recommendation that the company should 'postpone until the next decade its strategy of seeking long-term associations with foreign oil companies to develop its . . . reserves of extra-heavy crudes and instead "focus its efforts" . . . on exploration and development of its undiscovered light and medium crude reserves, which would be more cost-effective to produce' (*PON*, 28 January, 1993, p.1). PdVSA is probably reluctant to make the whole future of the Venezuelan oil sector contingent upon finding new reserves of light and medium crudes, although BP estimates that up to 70 per cent of the country's light and medium crude oil potential has yet to be tapped and believes that about 15 billion barrels of light and medium crude are waiting to be discovered in Venezuela.[7] Company officials, however, must still remember vividly the vast amounts of time and resources devoted to exploration for new oil from 1978 onwards, and the extremely poor results that they had to show for their efforts until 1985. In sum, it is logical to conclude that PdVSA is conscious of the acute need for new crude reserves, but this assumption must be qualified by the fact that the company is also painfully aware of two other facts: firstly, that exploring for oil in an oil province that has been active since 1917 is a very costly, risky, and protracted process and secondly, that past experience has conditioned Venezuelan policy-makers against putting all of their upstream eggs in one basket (such as the mammoth 30-year, $20 billion exploration programme recommended by the BP study in order to replenish Venezuelan reserves at a rate of 750 million barrels per year in the next decade). In short, BP should not take offence if PdVSA does not heed its recommendations to postpone oil belt development until 2010, because no matter what the British company thinks of its appraisals, the truth is that the only way to prove them right or wrong involves using the drill bit, and Venezuela is too dependent on oil to trust its future livelihood solely to the results of a drilling programme.

The BP study to which we have made continuous reference in this chapter also identifies possible policy options which PdVSA could follow in order to ensure its continued upstream well-being. It is instructive to review these options, in order to get a feel for the oil policy that may come out of Caracas in the years to come. The first option contemplates PdVSA funding its exploration and development programme on its own, with all the technological and financial risk that this implies. However, this course of action is effectively closed, because the Venezuelan government's confiscatory taxation policy (which has never served as a tool of economic rationality for PdVSA) puts too many financial constraints on the company. Of late, some encouraging noises – as far as PdVSA's finances are concerned – have been coming from the Venezuelan presidential palace: there is, for example, talk of a possible law that would limit (and eventually eliminate) the tax reference price for Venezuelan oil. However, as we have seen, the Venezuelan government's financial straits cast a shadow over this kind of initiative (for instance, Venezuela will have to start paying principal on the restructured portion of its $28.5 billion external debt in 1995–6).[8]

Venezuela's second option is for the company to pursue its 'strategic associations' schemes. As we have seen, the directors of foreign oil companies have greeted these offers with caution and not too much enthusiasm. Although they are not the type of entrepreneurs who would pass by an offer of equity crude without so much as a cursory glance at it – and that is why many of them have been signing agreements which, as of today, are no more than a statement of intentions – they are aware that many of their peers who have invested heavily in high-cost oil provinces 'are either losing money or have small prospects of making money unless assisted by an increase in the price of oil, which is not to be counted on' (*WPA*, 29 June, 1992, p.1). Although oil companies would ideally want to have larger amounts of equity crude reserves in their portfolio of assets, at this point in time they would rather that these reserves were located in low-cost areas, where investments would still yield reasonable returns should crude oil prices fall to the floor set by production costs in the Middle East. This would assure them profits even if the floor for crude oil's world price were to be set by the integrated state producers of the Middle East. Seen in this light, the prospects for a relatively high cost area like Venezuela might seem unpromising. However, oil companies remain wary of the historical trends which led to the great 1970s nationalizations and like to keep their options open. As Paul Frankel says:

> 'If the world were one unit . . . massive low-cost reserves [like those of the Middle Eastern countries] would set the tone of

the market to the virtual exclusion of higher-cost alternatives. But the world is not one unit and social and political considerations make unacceptable full dependence on these small areas of concentrated resource. Hence the need for finding, developing and maintaining higher-cost oil which has to enter the market alongside the low-cost variety' (Frankel, 1989, p.3).

Therefore, the probability that major oil companies will concentrate *solely* on developing oil resources in the Middle East (in the uncertain event that the countries in this region decide to open their oil sectors to foreign ownership again), is today as unlikely as ever. This may give Venezuela a window of opportunity to attract international capital to its oil sector. However, in order to succeed, it will have to compete with other relatively high-cost oil provinces, such as Russia, the North Sea or even the Central Asian republics of the CIS. The fact that oil reserves found in any of these places are of a more conventional nature than Orinoco reserves probably means that Venezuela has to lure the oil companies with something better than the promise of equity participation in the Orinoco and concessions in conventional fields in other parts of Venezuela. Nevertheless, PdVSA's ideal scenario was aptly summarized by one of its senior board members when he said: 'What we envision by the end of the century is a Venezuelan industry where PdVSA is competing . . . against international companies operating in the production and refining of heavy crudes, and also in the exploration and production of conventional crudes in high risk areas. There's no other way, PdVSA can't go it alone' (*PON*, 26 May, 1992, p.5).

The third option for PdVSA is 'to enter into contractual partnerships with foreign oil companies for exploration and development of light and medium crudes, with Venezuela maintaining control of the partnership' (*PON*, 28 January, 1993, p.4). Rather unsurprisingly, the opinion of BP (which is obviously interested in entering into equity agreements) is that 'this option', reminiscent of a straightforward stand-alone equity concession agreement, 'offers Venezuela the highest value in the near and long terms' (*ibid.*). Whether Venezuela will take this course of action will probably depend on the number of 'strategic alliances' for exploiting the Orinoco Belt that eventually materialize (so far, aware of the symbolic significance implicit in any backtracking regarding the equity issue, both the Venezuelan government and PdVSA have insisted that the granting of concessions for exploration and development of light and medium crude reserves should be contingent on a long-term commitment on the part of any prospective concessionaire to develop the reserves in the Orinoco Oil Belt). It should be remembered that, if the strategic alliances programme fails to come through, however, the whole future of petroleum development in Venezuela – as delineated in the 1991–2000 long-term plans – will be at risk. Therefore, in the event of such a failure occurring or appearing inevitable, the Venezuelan government might consider opening up the conventional hydrocarbons basins to foreign companies, in the hope that once they are established in the country, it will be able to persuade them to go into the Orinoco at a later date. As an old Ukrainian proverb says: 'You must sing the tune of the person on whose wagon you ride'. A final word is in order. The fact that the second and third options are being discussed at all is very significant, because it means that in the future the fate of the Venezuelan oil industry might become divorced from the fate of PdVSA, for the first time since the nationalization of Venezuelan oil. These options (along with reactivation of marginal oilfields or the Cristóbal Colón LNG project, to give two examples) represent a radical departure from the established practice regarding foreign (or private, for that matter) ownership of the nation's hydrocarbon resources. One need only remember, for instance, the public outcry that shook Venezuela when service contracts to develop the Orinoco Oil Belt were awarded to engineering firms Bechtel and Lummus (Coronel, 1983, p.185 *ff.*), to appreciate the chasm that divides present from past attitudes towards foreign investment in oil.[9] This change in public opinion has come at a rather convenient time for the Venezuelan government, because PdVSA's financial difficulties mean that the oil sector's reliance on foreign capital influx in the future may be quite substantial. Venezuela's new foreign investment law, embodied in Executive Decree 727, effective from 18 January, 1990, is a substantial improvement over the last one (Decree 1200 of 1986) and reflects these changing concerns.[10]

Nevertheless, ambiguous aspects remain which some potential foreign investors would undoubtedly prefer to see clarified before they actually put any money into an oil venture in Venezuela. A reduction in the high income tax rate on extraction and sales of crude oil is surely the most important among all of the contentious issues currently on the agenda. One should also ask whether the opening-up to foreign oil companies will raise the question of Venezuela's participation in OPEC. In many quarters OPEC membership has been seen as a potential hindrance for foreign investment, and the government has received proposals suggesting that a re-evaluation of the country's position in the organization is in order, because as some have said, even though 'there's tremendous international interest in getting back into Venezuela's oil sector . . . there can't be any type of quota restrictions on Venezuela' (*PON*, 26 May, 1992, p.5). However, foreign oil companies have long been active in a number of OPEC member countries – the UAE, Nigeria, Indonesia and even Libya, and they have recently been involved again in Algeria and Qatar. While Venezuela may feel constrained by OPEC's production policies, nevertheless for both revenue reasons and the economic feasibility of heavy crude oil development prospects, the country requires whatever boost OPEC may give, even in weak market conditions, to the oil price.

Notes

1. Indeed, it is my opinion that, had it not been granted a breather by the aforementioned discoveries, PdVSA would probably have needed to enlist the aid of foreign capital on a massive scale, had it wanted to continue developing Venezuela's hydrocarbons resources.
2. PdVSA's expertise in these areas is far superior to that of other third world national oil companies.
3. Indeed, it is hard to disagree with those who regard PdVSA as the epitome of corporate success at the state enterprise level. Even in circles opposed in principle to the existence of state-owned firms, it is recognised that the company has been the protagonist of a notable success story (see *The Economist*, 11 January, 1992, pp. 77–8).
4. Entitled *Venezuelan Production 1920-2030.*
5. According to the BP study referred to above, it may amount to $60

billion, with $10 billion earmarked for production infrastructure, and $50 billion for upgrading plants.

6. In this article, Adelman also quotes the example of the Kern River Field, discovered in 1899 and assessed to have 54 million barrels of reserves after 43 years in production. By 1985, the field had produced 736 million barrels (instead of 54 million), and it had remaining recoverable reserves of over 900 million barrels.
7. BP says that the most promising areas are the Orinoco River Delta platform, the Delta Amacuro region, the northeast part of Monagas state, the northwest part of Barinas state, the Paraguaná-Maturín region, the Andean North Mountain Flank, and the Perijá region to the west of Lake Maracaibo. According to BP, 70 percent of the possible reserves in these regions are concentrated in the Delta Amacuro region, the Orinoco delta, and the unexplored parts of Monagas and Barinas states.

   It should be pointed out that during PdVSA's 1978-86 exploration effort, Lagoven drilled in the Orinoco Delta region (which also covers part of the Delta Amacuro territory) with a marked lack of success (Coronel, 1983, p.131), and that BP's numbers are much more optimistic than the 1983 DOE/EIA study quoted earlier, which said there was a 95 per cent probability that 12 billion barrels of conventional oil sources remained to be discovered in Venezuela, and that the probability of finding 38 billion barrels of reserves was only 5 per cent.
8. Out of Venezuela's total debt, about $20 billion were subject to a Brady plan restructuring in 1990.
9. Venezuelan congressmen denounced the contracts because, in their eyes, they would deliver the Venezuelan oil sector into the hands of foreigners once again, barely five years after the country had rid itself of its concessionaires.
10. The new foreign investment legislation eliminates all limitations on dividend remittances for foreign companies, removes all limitations regarding the reinvestment of earnings and repatriation of capital and gives foreign companies equal access to Venezuelan credit (both medium and long term). The 2 per cent R&D withholding requirement disappears and the payment of inter-company royalties – from Venezuelan subsidiaries to foreign parent companies – in exchange for technical assistance, trademarks and royalties is allowed for the first time. It also enables foreign companies to form new fully controlled subsidiaries without prior consent or approval by SIEX (the Venezuelan authority overseeing foreign investment matters) and gives foreign investors the possibility of purchasing shares of Venezuelan companies in the stock exchange or from Venezuelan nationals, without obtaining prior written approval from SIEX (although notification of such purchase still has to be filed with SIEX). However, it has not lifted the equity restrictions affecting certain reserved sectors (like oil). Before this can happen, the Venezuelan congress has to approve PdVSA's joint venture proposals which have already been put up for its review and consideration (like Cristóbal Colón), or will be in the future (like the refining ventures in the Orinoco). However, PdVSA – through an intensive lobbying effort

– has already succeeded in eliminating many obstacles for foreign investment in oil, and this will undoubtedly enhance the possibility of large flows of international investment materializing if Congressional approval for PdVSA's proposals is forthcoming. For instance, PdVSA has made it clear that it will take no more than a 49 per cent share in any joint venture deal struck with a private company, whether in petrochemicals or oil and gas. Also, the tax rates on joint ventures in gas, refining and petrochemicals have been reduced dramatically.

APPENDIX 1: PROSPECTIVE VENEZUELAN OFFSHORE AREAS

The exploration of the Venezuelan continental platform began in 1947, with a magnetometric survey of the Gulf of Venezuela. Years later, many concessionaires, as well as CVP, began drilling for oil in the Gulf of Pariá. Between 1956 and 1960, not one of the 21 wells drilled found oil in sufficient quantities to justify development. Between 1978 and 1983, PdVSA's operating subsidiaries undertook another offshore exploration programme, which embraced all of the unexplored offshore basins except the Gulf of Venezuela. The company's estimates as to the quantity of oil which it hoped to find in each area are given in Table A.1.1

Table A.1.1: Venezuela. Estimate of Possible Oil Reserves in Unexplored Offshore Areas. Billions of Barrels.

Area	Billions of Barrels
Gulf of Venezuela	4.0
Onshore and offshore Orinoco Delta	4.0
Gulf of La Vela	0.3
Gulf of Pariá	0.3
South Lake Maracaibo	0.3
Offshore Puerto Cabello	0.2
Tuy-Cariaco Basin	0.3

Source: Coronel, 1983, p. 130.

The exploration programme proved somewhat disappointing. Maraven's results in the Golfo Triste area (offshore Puerto Cabello) indicated that the area had no oil potential whatsoever. The company, though, found small amounts of good quality oil in the Tuy-Cariaco basin. Lagoven found commercial accumulations of gas to the north of the Pariá peninsula, but it came up with nothing when it drilled in the Orinoco delta region. Finally, Corpoven's efforts in the Gulf of La Vela proved similarly futile. However, even in the face of this record, the recent proving of reserves in the Boquerón structure, as well as the gas discoveries in the Carúpano basin, seem to confirm the great hydrocarbon potential of other

Venezuelan offshore areas. Venezuelan hopes centre on the Gulf of Venezuela. Indeed, there are many who think that beneath its waters will be found oil fields that will overshadow those of the Campeche Sound (Coronel's estimate of 4 billion barrels of reserves would be seen by them as much too conservative).[1] In any case, if the border issue with Colombia is ever solved, and Venezuela starts drilling for oil in the Gulf of Venezuela, the first results of this drilling programme will send many petroleum engineers worldwide scrambling for their computers, in order to get a picture of the area's true potential.[2]

Notes

1. A 1983 estimate of the US Department of Energy was more pessimistic, however. According to the DOE, only 5 billion barrels of oil remained to be discovered in all of Venezuela's northern offshore (and onshore) basins: the Falcón and Tuy-Cariaco basins, the Gulf of Venezuela, the Bonaire basin and the Gulf of La Vela. The same study said that only 1.49 billion barrels of oil would be discovered in the Margarita-Tobago basin, which includes the Gulf of Pariá, and the Orinoco Delta zone (DOE/EIA, 1983, p.4).
2. Many studies have shown that the size/frequency distribution of fields in a sedimentary basin is lognormal. In other words, when the sizes of all the fields in a basin are plotted on a logarithmic scale against a cumulative per cent, the resulting curve approximates a straight line (see Selley, 1985, p. 417), in which the number of fields increases as the field size factor decreases. The slope of the fitted line varies from basin to basin, but once the slope for a given basin has been established (by finding a few fields), the ultimate recoverable reserves of the basin can be estimated, because when the minimum size of an economically viable field is known, so is the probability of finding a field this size.

APPENDIX 2: PdVSA'S 1991-3 MARGINAL OIL FIELD REACTIVATION PROGRAMME

Table A.2.1: Characteristics of the Oil Fields Offered in First Bidding Round.

Production unit	*Fields (no.)*	*Type of oil*	*Remaining reserves*		*Cumulative production+*
			Proved+	*Probable+*	
			Million Barrels		*Million Barrels*
Quiriquire	1	H*	55	115	757
Jusepín	1	L, M	32	89	194
South Monagas	3	H	78	19	19
Pedernales	1	H	49	185	59
West Falcón	4	L, M	2	-	50
North Falcón	2	L	10	362	58
Offshore Falcón	1	C, L, M	54	208	-
West Guárico	13	L, M, H	29	112	135
East Guárico	20	L, M	48	79	112
TOTAL	46		357	1,169	1,384

*C: Condensate; L: Light; M: Medium; H: Heavy
+In millions of barrels
Source: *O&GJ*, 19 August, 1991, p. 15

Table A.2.2.: Characteristics of the Oilfields Offered in the Second Bidding Round.

Production Unit	*Fields (no.)*	*Type of Oil*	*Remaining Proved Reserves Million Barrels*	*Cumulative Production + Million Barrels*
West Guárico	13	L, M, H*	29	135
Sanvi (Guere)	16	L	50	34
Oritupano-Leona	15	M	169	255
Quiamare-La Ceiba	6	L	119	71
Casma-Soledad	1	M, H	45	-
Jusepín + +	1	L, M	31	194
Quiriquire + +	1	H	94	757
West Urdaneta	1	L, M	110	32
West Falcón + +	4	L, M	2	50
North Falcón + +	2	L	10	57
Offshore Falcón + +	1	C, L, M	54	-
Colón	6	L	134	267
West Zulia	6	L, M, H	426	24

* C: Condensate; L: Light; M: Medium; H: Heavy
+ In millions of barrels
+ + These fields have been put on offer again, even though they received no bids in the first round.
Source: *O&GJ*, 7 December, 1992, p.35.

Table A.2.3: Bids Listed for Venezuela's Marginal Oil Fields, First Bidding Round.

Production unit	*Bidders*
South Monagas	Armstrong Petroleum Corp.
	Vinccler CA/Benton Oil & Gas
	Cementaciones Petroleras/Cía. Naviera
	Pérez Companc
	Inversiones Wineca
	Saskatchewan Oil & Gas Co.
Pedernales	BP Venezuela Ltd.
	Shell de Venezuela SA
North Falcón	Benton/Vinccler
	Welltech Inc./Inversiones Dromug CA/TD
	Instalaciones
	Lingoteras de Venezuela SA
	Creston Exploration Inc./Intera
	Technologies Corp./Ingeniería 5020 CA
West Guárico	Cie. Geofinanciere (Geoservices)
	Olympic Oil & Gas Corp.
	Instalaciones Lingoteras de Venezuela SA
East Guárico	Teikoku Oil Co. Ltd.
	Astra Capsa/Olympic Oil Co. Ltd.
	Cie. Geofinanciere
	Repsol Exploration
	Petróleos y Derivados Tamanaco S.A.

Source: *O&GJ*, 16 March, 1992, p. 36.

APPENDIX 3: ASSAYS OF SOME REPRESENTATIVE VENEZUELAN CRUDES

Bachaquero BCF 17

Crude
Gravity, °API: 16.8°
Sulphur, wt %: 2.4
Viscosity, SUS @ 100°F: 1,362
Pour point, °F: -10
Yield, 0°(IVT)-82°F, vol %:0.80

Light naphtha
Range, °F: 82–200
Yield, vol %: 2.44
Sulphur, wt %: 0.0075
P/N/A %: 66.17/31.28/2.55

Heavy Naphtha
Range, °F: 200–300
Yield, vol %: 3.54
Sulphur, wt %: 0.0266
P/N/A %: 29.38/58.96/11.66
RON clear: 66.7

Naphtha
Range, °F: 300–350
Yield, vol %: 1.81
P/N/A %: 23.20/59.59/17.21
Sulphur, wt %: 0.1107

Kerosene
Range, °F: 350–400
Yield, vol %: 2.11
Sulphur, wt %: 0.19
P/N/A %: 25.12/54.49/20.39

Freeze point, °F: -74.12
Smoke point, mm: 17.75
Cetane no.: 48.56

Gas oil
Range, °F: 400-500
Yield, vol %: 6.37
Gravity, °API: 35.43
Sulphur, wt %: 0.52
Aromatics, wt %: 25.29
Freeze point, °F: -66.62
Smoke point, mm: 9.23
Cetane no.: 46.36
Aniline point:, °F: 127.39
Cetane index: 39.98

Gas oil
Range, °F: 500–550
Yield, vol %: 3.93

Gas oil
Range, °F: 550–650
Yield, vol %: 8.95
Sulphur, wt %: 1.36
Smoke point, mm: 2.83
Pour point, °F: -47.99
Aniline point:, °F: 136.64

Residue
Range, °F: 650+
Yield, vol %: 70.24
Gravity, °API: 9.39
Sulphur, wt %: 3.0
Kinematic viscosity @ 122° F: 8,672
Ni/V, ppm: 74/435

Boscán

Crude
Gravity, °API:10.1
Sulphur, wt %: 5.5
Viscosity, SUS @ 100°F: 90,000
Pour point, °F: +50

Light naphtha
Range, °F: 82–200
Yield, vol %: 0.55

Heavy Naphtha
Range, °F: 200–300
Yield, vol %: 1.32
Sulphur, wt %: 0.068
Paraffins: 53.28

Naphtha
Range, °F: 300–350
Yield, vol %: 0.88
Sulphur, wt %: 1.28
Paraffins %: 34.9

Kerosene
Range, °F: 350–400
Yield, vol %: 1.28
Sulphur, wt %: 1.88
Aromatics %: 24.04
Freeze point, °F: -141.0
Smoke point, mm: 16
Aniline point:, °F: 112.69

Gas oil
Range, °F: 400–500
Yield, vol %: 3.96
Gravity, °API: 34.0
Sulphur, wt %: 3.14
Kinematic viscosity @ 122° F: 1.81
Freeze point, °F: -77.53

Smoke point, mm: 16.11
Aniline point:, °F: 113.83
Cetane index: 37.24

Gas oil
Range, °F: 500–550
Yield, vol %: 2.66
Sulphur, wt %: 4.02

Gas oil
Range, °F: 550–650
Yield, vol %: 6.46
Gravity, °API: 25.54
Sulphur, wt %: 4.47
Pour point, °F: -10.9
Aniline point:, °F: 117.3
Cetane no.: 38.0
Cetane index: 42.27

Residue
Range, °F: 650+
Yield, vol %: 82.9
Gravity, °API: 6.10
Sulphur, wt %: 5.86
Kinematic viscosity @ 122° F: 22,502
Ni/V, ppm: 175/1,407

Ceuta

Crude
Gravity, °API: 31.8°
Sulphur, wt: 1.2%
Viscosity, SUS @ 100°F: 62
Pour point, °F: -35
Yield, 0(IVT)-82°F, vol %:2.10

Light naphtha
Range, °F: 82–200
Yield, vol %: 7.51
Sulphur, wt %: 0.002
P/N/A %: 75.69/19.85/4.46

Heavy Naphtha
Range, °F: 200–300
Yield, vol %: 9.89
Sulphur, wt %: 0.01
P/N/A %: 50.88/32.07/17.05
RON clear: 71.5

Naphtha
Range, °F: 300–350
Yield, vol %: 5.36
P/N/A %: 42.25/38.78/18.98

Kerosene
Range, °F: 350–400
Yield, vol %: 5.28
Sulphur, wt %: 0.054
P/N/A %: 39.8/41.1/19.0
Smoke point, mm: 21.39
Aniline point:, °F: 122.8
Cetane index: 40.9

Gas oil
Range, °F: 400–500
Yield, vol %: 9.93
Gravity, °API: 38.75
Sulphur, wt %: 0.33
Aromatics, %: 19.10
Smoke point, mm: 19.33
Aniline point:, °F: 136.18
Cetane index: 41.84

Gas oil
Range, °F: 500–550
Yield, vol %: 5.0
Sulphur, wt %: 0.72

Gas oil
Range, °F: 550–650
Yield, vol %: 9.24
Gravity, °API: 30.17

Sulphur, wt %: 1.0
Aniline point:, °F: 155.8
Cetane index: 48.03

Residue
Range, °F: 650+
Yield, vol %: 45.69
Gravity, °API: 16.60
Sulphur, wt %: 2.05
Ni/V, ppm: 54/281

Tía Juana Light

Crude
Gravity, °API: 32.1°
Sulphur, wt %: 1.1
Pour point, °F: -45
Yield, 0(IVT)-82°F, vol %:2.16

Light naphtha
Range, °F: 82–200
Yield, vol %: 5.19

Heavy Naphtha
Range, °F: 200–300
Yield, vol %: 10.82
Sulphur, wt %: 0.0017
P/N/A %: 65.99/24.30/9.71

Naphtha
Range, °F: 300–350
Yield, vol %: 4.82
P/N/A %: 51.23/31.69/17.08
Sulphur, wt %: 0.0077

Kerosene
Range, °F: 350–400
Yield, vol %: 4.64
Sulphur, wt %: 0.0281
P/N/A %: 45.77/35.64/18.58

Smoke point, mm: 23.37
Aniline point: 134.99

Gas oil
Range, °F: 400–500
Yield, vol %: 10.41
Gravity, °API: 40.72
Sulphur, wt %: 0.11
Smoke point, mm: 14.99
Aniline point:, °F: 134.99
Cetane index: 50.43

Gas oil
Range, °F: 500–550
Yield, vol %: 4.96
Sulphur, wt %: 0.4540
Smoke point, mm: 11.99
Aniline point:, °F: 158.47

Gas oil
Range, °F: 550–650
Yield, vol %: 9.25
Sulphur, wt %: 0.78
Smoke point, mm: 14.42
Aniline point:, °F: 168.05

Residue
Range, °F: 650+
Yield, vol %: 47.75
Gravity, °API: 16.46
Sulphur, wt %: 1.87
Ni/V, ppm: 26/224

BIBLIOGRAPHY

BOOKS, ARTICLES AND GOVERNMENT PUBLICATIONS

Adelman, M. A. (1972), *The World Petroleum Market*, Baltimore: The Johns Hopkins University Press.

— (1986), *Oil Producing Countries' Discount Rates*. MIT Energy Laboratory Working Paper MIT-EL 86-015 WP. Cambridge, Mass.: MIT.

— (1989), 'Mideast Governments and the Oil Price Prospect', *The Energy Journal*', vol. 10, no.2, pp. 15–24.

Ako, B.D., A.O. Alabi, O.S. Adegoke and E.I. Enu (1983), 'Application of Resistivity Sounding on the Exploration for Nigerian Tarsand', *Energy Exploration & Exploitation*, vol. 2, no. 2, pp. 155–64.

Allison, Graham T. (1971), *Essence of Decision. Explaining the Cuban Missile Crisis*, Boston: Little, Brown and Company.

American Petroleum Institute (API) (a) 1970–1992 *Basic Petroleum Data Book. Petroleum Industry Statistics*, Washington D.C.: API Finance Accounting and Statistics Department. (b) 1985–1992 *Imported Crude Oil & Petroleum Products*, Washington D.C.: API Finance Accounting and Statistics Department.

Amuzegar, Jahangir (1982), 'Oil Wealth: a Very Mixed Blessing', *Foreign Affairs*, vol. 60, no. 4, pp. 814–34.

Baker, George (1992),'Reserves Dispute Points to Need for US-Mexico Cooperation on Oil E&D', *Oil&Gas Journal*, vol. 90, no. 10, pp. 34–8.

Betancourt, Rómulo (1978), *Venezuela's Oil*, London: George Allen & Unwin.

Boy de la Tour, X., J.L. Gadon and J.J. Lacour (1986), 'New Oil: What's in the Future?', *Energy Exploration & Exploitation*, vol. 4, no.6, pp. 401–95.

The British Petroleum Company (1975–1991), *BP Statistical Review of World Energy*.

Brown, Keith C. (1989), 'Reserves and Reserve Production

Ratios in Imperfect Markets', *The Energy Journal*, vol. 10, no. 2, pp. 177–86.
CEPET (1989), *La industria venezolana de los hidrocarburos*, 2 vols, Caracas: Ediciones del Centro de Formación y Adiestramiento de Petróleos de Venezuela y sus Filiales.
Corden, W. M. (1982), 'Booming Sector and Dutch Disease Economics: Survey and Consolidation', *Oxford Economic Papers*, vol. 34, no. 3, pp. 359–80.
Corden, W. M. and J. P. Neary (1982), 'Booming Sector and De-Industrialisation in a Small Open Economy', *The Economic Journal*, vol. 92, pp. 825-48.
Coronel, Gustavo (1983), *The Nationalization of the Venezuelan Oil Industry. From Technocratic Success to Political Failure*, Lexington, Mass.: Lexington Books.
Danielsen, Albert L. (1982), *The Evolution of OPEC*, New York: Harcourt, Brace, Jovanovich.
Demaison, G.J. (1978), 'Tar sands and Super Giant Oil Fields', *Oil&Gas Journal*, vol. 77, pp. 202–5.
Department of Energy, Energy Information Administration (DOE/EIA) (1983), *The Petroleum Resources of Venezuela and Trindad and Tobago*, Washington D.C.: National Energy Information Center.
— (1983a), *The Petroleum Resources of Mexico*, Washington D.C.: National Energy Information Center.
di Filippo, Armando (1981), *Desarrollo y desigualdad social en América Latina*, México, Fondo de Cultura Económica.
Echeverría, Oscar A. (1985), *La economía venezolana 1944–1985*, Caracas: Federación de Cámaras y Asociaciones de Comercio y Producción (FEDECAMARAS)
Evans, John and Gavin Brown (eds) (1991), *OPEC and the World Energy Market. A Comprehensive Reference Guide*. Harlow: Longman.
Frankel, Paul (1989), 'Principles of Petroleum – Then and Now', *The Energy Journal*, vol. 10, no. 2, pp. 1–5.
Gelb, Alan, *et. al.* (1988), *Oil Windfalls. Blessing or Curse?* Oxford: The World Bank.
Good, Barry C. (1985),'Memorandum on Imperial Oil Limited', Morgan Stanley International Investment Research.
Hausmann, Ricardo (1990a), *Dealing with Negative Oil Shocks: the Venezuelan Experience in the Eighties*, (Mimeo).

— (1990b) *Shocks externos y ajuste macroeconómico*, Caracas: Banco Central de Venezuela.

Hirschman, Albert O. (1981), *Essays in Trespassing: Economics to Politics and Beyond*, New York: Cambridge University Press.

International Monetary Fund (IMF)(1948–1992), *International Financial Statistics*, Washington D.C.: IMF.

Lieuwen, Edwin (1985), 'The Politics of Energy in Venezuela', in *Latin American Oil Companies and the Politics of Energy*. John D. Wirth (ed.) Lincoln: University of Nebraska Press, pp. 189–225.

Little, Ian, Tibor Scitovsky and Maurice Scott (1970), *Industry and Trade in Some Developing Countries. A Comparative Study*, Oxford University Press: O.E.C.D. Development Centre.

Mabro, Robert (1989),*OPEC's Production Policies. How Do They Work? Why Don't They Work?* Oxford: Oxford Institute for Energy Studies.

El Mallakh, Ragai, (ed.) (1983), *Heavy Versus Light Oil: Technical Issues and Economic Considerations*. Boulder, Colorado: International Research Center for Energy and Economic Development (ICEED).

Martínez, Aníbal R. (1980), *Gumersindo Torres. The Pioneer of Venezuelan Petroleum Policy*, Caracas: PdVSA.

— (1986), *The Journey from Petrolia*, Caracas: Edreca.

— (1987), 'The Orinoco Oil Belt, Venezuela', *Journal of Petroleum Geology*, no. 2, pp. 125–34.

— (1989), *Venezuelan Oil: Development and Chronology*, London: Elsevier Applied Science.

Mayobre, Eduardo (1987), 'La renegociación de la dueda externa de Venezuela en 1982–1983', in *La crisis de la dueda externa en América Latina*, Miguel S. Wionszek (ed.). vol. 2, Mexico.

McGowan, Francis (1990), 'The Development of Orimulsion and Venezuelan Oil Strategy', *Energy Policy*, vol. 18, no. 10, pp. 913–26.

Meyer, Richard F. (1987), 'Prospects for Heavy Crude Oil Development', *Energy Exploration & Exploitation*, vol. 5, no. 1, pp. 27–55.

Millard, Vernon (1983),'Outlook for Canadian Heavy Crude Oil', in El Mallakh, *op. cit.*, pp. 163–86.

Ministerio de Energía y Minas de la República de Venezuela

(MEM) (1990), *Petróleo y otros datos estadísticos,* Caracas: MEM.
Mommer, Bernard (1989), 'Es posible una política petrolera no rentista?', *Revista del Banco Central de Venezuela*, vol. 4, no. 3.
Nehring, Richard (1980), 'The Outlook for World Oil Resources', *Oil & Gas Journal*, vol. 79, pp. 170–5.
Niering, Frank E. (1980), 'Lack of Firm Policy on Oil Sands', *Petroleum Economist*, vol. 47, no. 2, pp. 65–71.
— (1982),'Doubts over Long-term Objectives', *Petroleum Economist*, vol. 49, no.3, pp. 85–7.
— (1989), 'Venezuela. Change of Direction in Oil Policy', *Petroleum Economist*, vol. 56, no. 10, pp. 218–21.
Noreng, Oystein (1980), *The Oil Industry and Government Strategy in the North Sea*, Boulder: ICEED.
Obadan, Michael I. (1986), 'The Impact of Export Instability on the Economic Development of Nigeria: a Statistical Verification', *OPEC Review*, vol. X, no. 4, pp. 409–26.
Ochoa, Orlando (1992), 'The Changing Politics of Venezuela in 1992: Oil, Democracy and Economic Reforms', *Oxford International Review*, June, pp. 22–4.
Odell, Peter (1986), *Oil and World Power*, Harmondsworth: Pelican Books.
Organization of Petroleum Exporting Countries (OPEC) (1970–1990), *OPEC Annual Statistical Bulletin*, Vienna: OPEC Secretariat.
OPEC* (1984), *Petroleum Product Prices and Their Components in Selected Countries. 1980–1983,* Vienna: OPEC Secretariat.
— (1991), *Petroleum Product Prices and Their Components in Selected Countries. 1983–1990*, Vienna: OPEC Secretariat.
Peebles, Malcolm W.H. (1992), *Natural Gas Fundamentals*, Bath: Shell International Gas.
Petróleos de Venezuela S.A. (PdVSA) (1976–1991), *Annual Report*, Caracas: PdVSA.
Philip, George (1982), *Oil and Politics in Latin America. Nationalist Movements and State Companies*, Cambridge: Cambridge University Press.
Randol, William L., and Ellen Macready (1987), 'Memorandum on Imperial Oil Limited', First Boston Equity Research.
Reader, Carol (1992), 'Venezuelan Fuel Advances Across the World', *Petroleum Review*, vol. 46 no. 540, pp. 30–32.

Riva, Joseph P. (1991), 'Dominant Middle East Oil Reserves Critically Important to World Supply', *Oil & Gas Journal*, vol. 89, no. 48, pp.62–8.

Schneider, Steven A. (1983), *The Oil Price Revolution*, Baltimore: Johns Hopkins.

Selley, Richard C. (1985), *Elements of Petroleum Geology*, New York: W.H. Freeman.

Servello, Juan (1983), 'Heavy Oil Development and the Venezuelan Petroleum Industry: Technical Issues and Economic Considerations' in El Mallakh, *op. cit.*

Seymour, Adam (1992), *Refining and Reformulation: The Challenge of Green Motor Fuels*, Oxford: Oxford Institute for Energy Studies.

Sterner, Thomas (1989),'Oil Products in Latin America: The Politics of Energy Pricing', *The Energy Journal*, vol. 10, no. 2, pp. 25–46.

Tiratsoo, E.N. (1984), *Oilfields of the World*, Beaconsfield: Scientific Press.

Tugwell, Franklin (1975), *The Politics of Oil in Venezuela*, Stanford: Stanford University Press.

Wennekers, J.H.N., *et. al.* (1979), 'Heavy Oil, Tar Sands Play Key role in Alberta, Saskatchewan Production', *Oil&Gas Journal*, vol. 78, pp. 291–304.

Werz, Nikolaus (1990),'State, Oil and Capital Accumulation in Venezuela', in Christian Anglade and Carlos Fortín, *The State and Capital Accumulation in Latin America*, London: n.p., pp. 182–210.

World Petroleum Congress (1984), *Classification and Nomenclature Systems for Petroleum and Petroleum Reserves*, New York: John Wiley & Sons.

Yáñez Betancourt, Leopoldo (1987), 'La economía venezolana. Problemas y perspectivas', *El Trimestre Económico*, vol. 54 no. 216, pp. 727–68.

Yergin, Daniel, 1991, *The Prize. The Epic Quest for Oil, Money and Power*, New York: Simon and Schuster.

Zlatnar, Mirjana (1986), 'Venezuela Is Ready for Future Challenges', *OPEC Bulletin*, vol. xvi, no. 10, pp. 28-36. vol. xvi, no. 10, pp. 28–36.

JOURNALS, ANNUALS AND ENCYCLOPEDIAS

Bloomberg Oil Buyers' Guide
The Economist
Energy Economist (EE)
Euroil
Global Oil Report (GOR)
International Gas Report (IGR)
International Petroleum Encyclopedia (IPE)
Jet Fuel Intelligence (JFI)
Latin American Weekly Report (LAWR)
Oil Daily Energy Compass (EC)
Oil&Gas Journal (O&GJ)
Oil&Gas Journal Data Book
Oilweek
OPEC Bulletin (OB)
PDVSA Contact (PC)
Platt's Oilgram News (PON)
Petroleum Economist (PE)
The Petroleum Industry Indicators (PII)
Petroleum Intelligence Weekly (PIW)
Petroleum Press Service (PPS)
Petroleum Times (PT)
Shell World
US Oil Week
Weekly Petroleum Argus (WPA)
World Gas Intelligence (WGI)

INDEX